UPPER LEFT CITIES

UPPER LEFT CITIES

A CULTURAL ATLAS of SAN FRANCISCO, PORTLAND, and SEATTLE

HUNTER SHOBE AND DAVID BANIS
WITH ZURIEL VAN BELLE

Lead Contributors
Geoff Gibson and Sachi Arakawa

SASQUATCH BOOKS

SEATTLE

CONTENTS

III. SOCIAL RELATIONS

IV. COMMERCE

V. POPULAR CULTURE

PREFACE

This book grew out of our previous title, *Portlandness: A Cultural Atlas*, in which we concentrated exclusively on the city where we live. People asked us if there would be a sequel. Another book about Portland? Maybe one about other cities? We decided to do both.

After spending years mapping and writing about one city, we resolved to compare cities. We were curious about how the big cities on the West Coast were alike and how they differed. Initially we wanted to include Oakland, California, and Vancouver, BC, but we decided to focus our efforts on San Francisco, Portland, and Seattle, the major West Coast cities in the northern part of the United States—or what we call the Upper Left.

We attempt to blend academic and popular styles, which is a difficult balance to strike. We hope to bring academic research to people who don't usually read geography and urban studies journals, and bring storytelling and graphics to people who do. Each of us has personal connections to these cities.

Hunter traces his Upper Left connections to the summer of 1994. That year he had tickets to World Cup games at Stanford Stadium and the Rose Bowl. So he drove from Washington, DC, and spent the summer traveling around California, Oregon, and Washington with a couple of friends. Just before the Nevada–California border, Hunter made his buddies pull over on the shoulder of the highway so that he could walk, rather than drive, across the state line to commemorate his first time in California.

Hunter visited San Francisco, Portland, and Seattle on this trip. A few months later, he moved to San Francisco, where he lived until the fall of 1998. Ever since, he has returned for days, weeks, and months at a time. From San Francisco, Hunter moved to Eugene, Oregon, for graduate school. He began to visit friends in Seattle every year. In the summer of 2006, he moved to Portland to teach at Portland State University. The Upper Left cities became personally stitched together when he met his wife (who grew up north of Seattle) on a blind date in Portland set up by mutual friends from San Francisco.

In researching this book, Hunter covered hundreds of miles (and wore through several pairs of Sambas) walking across the three cities. The initial idea for an atlas comparing multiple cities came to him on his four-mile walking commute to campus.

David spent his formative years in Southern California, but he has called the Upper Left home ever since. He went to college at UC Berkeley and spent his early working life in the Bay Area, and continues to visit family there regularly. An engineering job at Boeing led him to Seattle, just as the *New York Times* began fawning over the city, Starbucks went corporate, and the grunge era dawned.

David was drawn to Portland for graduate studies in geography at Portland State University, again just as the rest of the country was discovering the city. He managed to stick around the Portland State geography department after grad school, where he teaches courses on cartography and geographic information systems. With a population of almost 2.5 million people, Portland is the smallest metro area in which David has lived. He enjoys the charming small-town feel of inner-city Portland, while at the same time loving the energy and intensity of Seattle and the Bay Area. Like many a confirmed urbanite, he escapes the city to the mountains, or some other place without many people, every chance he gets.

Collaborator Zuriel has lived in the orbits of San Francisco, Portland, and Seattle for most of her life. She grew up in the Blue Mountains of northeastern Oregon, about 245 miles away from Portland, one of the closest major cities. Seattle, though slightly farther afield, formed more of Zuriel's earliest impressions of city life. She took family trips to Seattle, where she visited her uncle and developed an affinity for bumptious gulls, endless drizzle, and Chinese seafood soups.

In adolescence, Zuriel moved from a town of about 650 people in rural Oregon to the bustling Bay Area, replacing pine trees with parking lots. Zuriel met her husband, Jonathan, while attending UC Berkeley. After graduation, and in search of a city with less bustle, more pine trees, and gloomier winters, they settled in Portland, where they have been ever since.

Everyone who worked on this atlas has deep connections to one or more of these cities. The result is that many perspectives are mixed into and shaped this book. While knowing that we could never include everything, through collaboration we hoped to explore as many features and connections of these cities as possible.

This is a book about San Francisco, Portland, and Seattle, but also a book about cities in general. We invite readers to reconsider their own blocks, neighborhoods, and cities and their many and varied connections.

This book came together with the contributions of students, alums, and colleagues at Portland State University, Portland Community College, and beyond. We all hold a deep respect for San Francisco, Portland, and Seattle, as well as for the people who live and work there. The responsibility of representing places in these three cities is one we took very seriously, even when adopting more playful themes and tones. Our goal was to display respect on every page.

Our team spent several years making this atlas, completing the final proofs in March 2020. We were on track for a fall 2020 release (or so we thought). The book was done. A few weeks later, COVID-19 upended lives throughout the United States, as it already had in places throughout the world. So much changed. This included our publication schedule which called for the book's release to be delayed a year.

By the fall of 2020, we realized that the book was not done. It needed updates (the election pages, wildfires) and the addition of new topics (COVID-19 and protests). Forced to work quickly we made the additions and revisions as best we could. Most of the book is a snapshot of these cities just before the coronavirus struck. We hope this serves as a reference point for considering what has changed since.

Whether you have lived in all three cities or have never visited any of them, we hope this book inspires you to reimagine San Francisco, Portland, and Seattle—the Upper Left.

INTRODUCTION: UPPER LEFT

This atlas explores San Francisco, Portland, and Seattle through maps, graphics, essays, and photos. While not exhaustive, this is a wide-ranging collection of stories about these cities and how they compare, contrast, and connect. This is not a guidebook, but rather a guide to thinking geographically and creatively about places and how they change.

Why does this atlas focus on these three cities? Although each is distinct, they also share similarities and are linked through shared subcultures. If you do a quick search for the greenest, most sustainable, transit-friendly, bike-friendly, healthy, exciting, or promiscuous cities, or for the best cities for hipsters, cats, dogs, coffee, twenty-somethings, or foodies, all three of these cities are usually listed in the top ten.

These West Coast cities are sometimes referred to as the Left Coast—an unofficial cultural region referenced in various media. Left Coast has carried a variety of meanings over the past thirty years, although it most often signals the liberalism shared among many of its inhabitants. In the national imagination, the Left Coast comprises the entire US West Coast, including Los Angeles; "liberal Hollywood" is quintessential Left Coast. While it captures many of the common cultural characteristics of San Francisco, Portland, and Seattle, the Left Coast encompasses a broader region.

In this atlas, we place San Francisco, Portland, and Seattle in a more specific subsection of the Left Coast—the Upper Left. We think the shared cultural identities of these three cities and their surrounding areas deserve a special designation. The Upper Left is an emerging term, a fuzzy concept still coming into view. With this atlas, we hope to make this term more distinct.

San Francisco, Portland, and Seattle have each been the "it" city. Each city was launched into the popular imagination of the country at different times—San Francisco during the late 1960s and 1970s, Seattle during the 1990s, and Portland during the 2010s. The popularity of each city caused rents and home prices to rise and some neighborhoods to change rapidly.

The three cities are linked physically through several different transportation networks. A single highway nearly connects them. Driving from Seattle to Portland to San Francisco is largely a matter of sitting in Seattle traffic long enough to find I-5 South and following it for 750 miles before merging onto I-80 and sitting in Bay Area gridlock on the way to the Bay Bridge.

Before the pandemic, thirty-five flights connected Seattle and the Bay Area, twenty-three flights connected Portland and the Bay Area, and some dozen flights connected Seattle and Portland. The Amtrak Coast Starlight connects Seattle, Portland, and Oakland before continuing on to Los Angeles.

To go by boat is to travel the route that first connected the three Upper Left cities. Portland, located at the confluence of the Columbia and Willamette Rivers, is sixty miles by boat to the Pacific Ocean. Seattle, situated around Elliott Bay, connects to the Pacific by way of the Salish Sea. San Francisco Bay connects directly to the Pacific Ocean via the Golden Gate strait.

The ecosystems and climates of each city, although distinct, are similar too. In Portland and Seattle, it drizzles a lot for most of the year. San Francisco gets less rain and more fog. Seattle temperatures are a little colder. These are temperate places with occasional weather extremes.

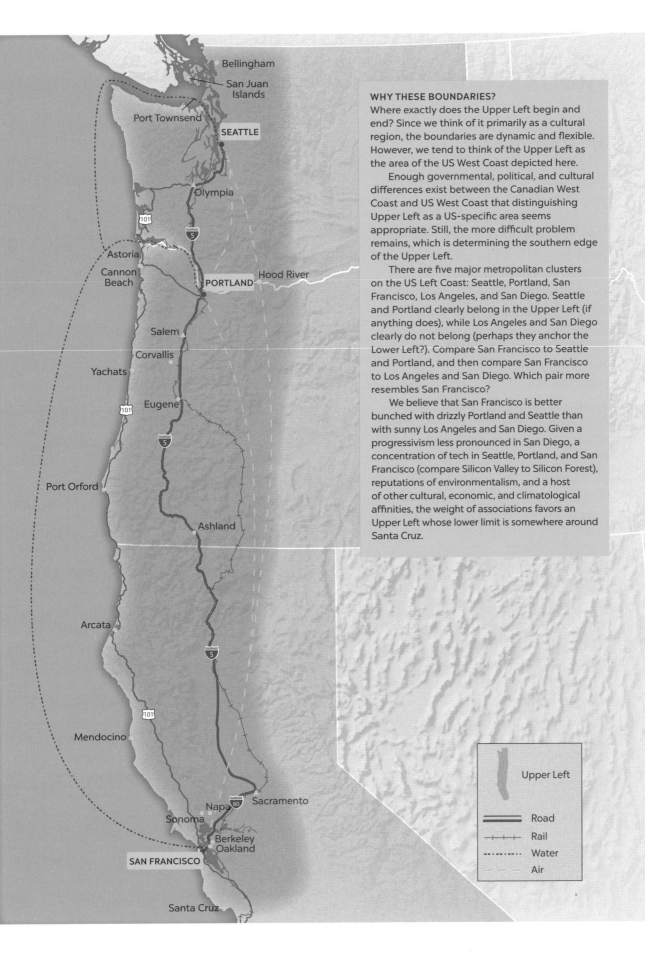

Bellingham

San Juan
Islands

Port Townsend

SEATTLE

Olympia

[101]

[5]

Astoria

Cannon
Beach

Hood River

PORTLAND

Salem

Corvallis

Yachats

[101]

Eugene

[5]

Port Orford

Ashland

Arcata

[5]

[101]

Mendocino

Napa [80] Sacramento

Sonoma

Berkeley
Oakland

SAN FRANCISCO

Santa Cruz

WHY THESE BOUNDARIES?

Where exactly does the Upper Left begin and
end? Since we think of it primarily as a cultural
region, the boundaries are dynamic and flexible.
However, we tend to think of the Upper Left as
the area of the US West Coast depicted here.

Enough governmental, political, and cultural
differences exist between the Canadian West
Coast and US West Coast that distinguishing
Upper Left as a US-specific area seems
appropriate. Still, the more difficult problem
remains, which is determining the southern edge
of the Upper Left.

There are five major metropolitan clusters
on the US Left Coast: Seattle, Portland, San
Francisco, Los Angeles, and San Diego. Seattle
and Portland clearly belong in the Upper Left (if
anything does), while Los Angeles and San Diego
clearly do not belong (perhaps they anchor the
Lower Left?). Compare San Francisco to Seattle
and Portland, and then compare San Francisco
to Los Angeles and San Diego. Which pair more
resembles San Francisco?

We believe that San Francisco is better
bunched with drizzly Portland and Seattle than
with sunny Los Angeles and San Diego. Given a
progressivism less pronounced in San Diego, a
concentration of tech in Seattle, Portland, and San
Francisco (compare Silicon Valley to Silicon Forest),
reputations of environmentalism, and a host
of other cultural, economic, and climatological
affinities, the weight of associations favors an
Upper Left whose lower limit is somewhere around
Santa Cruz.

Upper Left

———— Road

+++++ Rail

·········· Water

— — — Air

BACK IN THE DAY

SAN FRANCISCO

Before contact with Europeans, thousands of Ohlone lived in the Bay Area, including around the areas later known as Lake Merced and Mission Creek. In 1769, Europeans first caught sight of the bay during Gaspar de Portolá's overland expedition near present-day Pacifica.

In 1776, Spanish colonists arrived and founded the Presidio and Mission San Francisco de Asís (or Mission Dolores). The Mission, under the direction of Franciscan priest Junípero Serra, was named for Saint Francis of Assisi. The Mission chapel, still standing today, was built in 1791. The Spanish brought a new religion and an alien way of life to the native people, as well as the foreign disease of measles, which decimated the Ohlone and other native groups.

In 1821, Mexico declared itself independent from Spain, so San Francisco became part of Mexico. In the late 1830s, the slowly growing settlement was named Yerba Buena for a local plant. In 1847, to prevent an East Bay city from being named Francisca, the city was renamed San Francisco. That year, Irish engineer and surveyor Jasper O'Farrell platted a city plan with a diagonal double-wide boulevard that broke up the grid. Thus north–south streets rarely intersect at Market Street, causing daily traffic debacles today.

The biggest catalyst to San Francisco's growth occurred outside of the city. In 1847, Swiss native John Sutter charged carpenter James Marshall with finding a location in the foothills of the Sierra Nevada for a lumber mill to serve San Francisco's need for building materials. On January 24, 1848, Marshall found a small gold nugget in the mill's tailrace. Upon hearing the news most San Franciscans promptly rushed to the hills in search of gold.

Word soon spread worldwide. In the first year about ten thousand miners flocked to Northern California in search of gold. In 1849, ten times that number came. Yerba Buena Cove became littered with abandoned ships, some of which are buried under the financial district today. Less than two years after gold was found, San Francisco's population jumped from five hundred to twenty thousand. The city, as well as the rest of California, had become part of the United States in 1848 at the conclusion of the Mexican-American War.

In 1859, silver was discovered in western Nevada. The Comstock Lode provided a second boost of wealth and investment in San Francisco through the 1870s, during which thousands of new buildings were constructed.

The Central Pacific Railroad, which connected San Francisco to the East Coast, was completed in 1869. Chinese immigrants were the major labor force for West Coast railroad construction. Seeking economic opportunity, they were often subject to racist treatment. When the railroad was completed, many Chinese workers settled in San Francisco.

The last decades of the eighteen hundreds also saw the establishment of more than one hundred miles of cable car lines and the construction of Golden Gate Park, site of the California Midwinter International Exposition of 1894. As the map on the next page shows, at this time, urban development was minimal in the western half of the city.

On April 19, 1906, an estimated 7.7-magnitude earthquake shook San Francisco. After three days of fires, large parts of the city had been destroyed. Despite the massive losses, reconstruction of the city was rapid and most of downtown was rebuilt within three years.

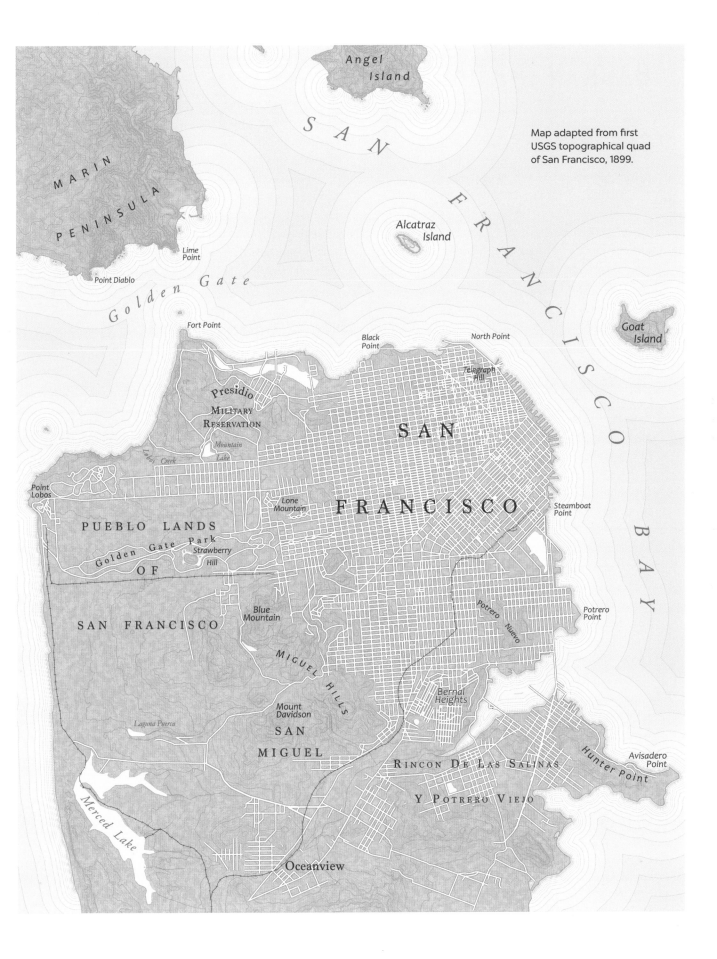

Map adapted from first USGS topographical quad of San Francisco, 1899.

Angel Island

SAN

Alcatraz Island

FRANCISCO

MARIN

PENINSULA

Lime Point

Point Diablo

Golden Gate

Goat Island

BAY

Fort Point

Black Point

North Point

Telegraph Hill

Presidio
MILITARY
RESERVATION

SAN

Mountain Lake

Lobos Creek

Point Lobos

Lone Mountain

FRANCISCO

Steamboat Point

PUEBLO LANDS

Golden Gate Park

Strawberry Hill

OF

SAN FRANCISCO

Blue Mountain

Potrero Nuevo

Potrero Point

MIGUEL HILLS

Mount Davidson

Bernal Heights

Laguna Puerca

SAN

MIGUEL

RINCON DE LAS SALINAS

Hunter Point

Avisadero Point

Merced Lake

Y POTRERO VIEJO

Oceanview

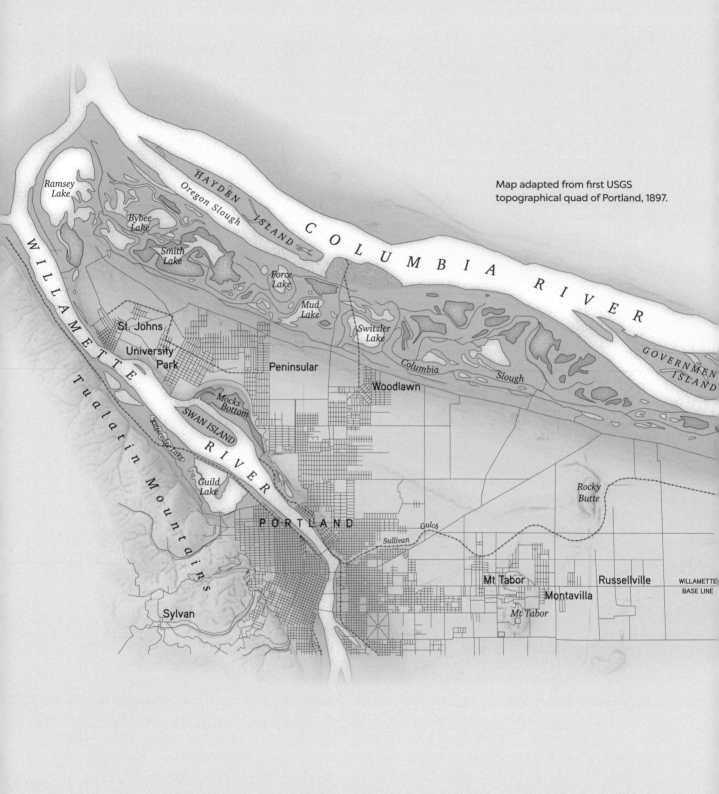

Map adapted from first USGS topographical quad of Portland, 1897.

COLUMBIA RIVER

WILLAMETTE RIVER

Tualatin Mountains

HAYDEN ISLAND

Oregon Slough

Ramsey Lake

Bybee Lake

Smith Lake

Force Lake

Mud Lake

Switzler Lake

Columbia Slough

GOVERNMENT ISLAND

St. Johns

University Park

Peninsular

Woodlawn

Mocks Bottom

SWAN ISLAND

Kittredge Lake

Guild Lake

PORTLAND

Sullivan Gulch

Rocky Butte

Sylvan

Mt Tabor

Mt Tabor

Montavilla

Russellville

WILLAMETTE BASE LINE

PORTLAND

Native Americans lived for thousands of years in what many now call the Willamette and Columbia Valleys. When Europeans arrived thousands of mostly Chinook-speaking peoples lived in the area, including the Multnomah who lived on what is now referred to as Sauvie Island. In the 1830s, less than forty years later, diseases brought by Europeans reduced native populations to a few hundred people.

Having missed the entrance to the Willamette River on two previous occasions, a group led by William Clark (of Lewis and Clark) encountered it in 1806. They navigated the Willamette from the confluence with the Columbia River as far south as the current University of Portland site in North Portland.

By 1843, Oregon City had emerged as an early contender for the key port location on the Willamette. The site that is now Portland, by contrast, was a tiny clearing in a densely wooded area west of a kink in the river with no permanent residents. Itinerant William Overton and lawyer Asa Lovejoy paid a fifty-cent fee to claim title to the 640 acres around the clearing that would soon be developed into downtown Portland. In 1844, Overton moved to California and sold his half of the claim to Oregon City merchant Francis Pettygrove.

In 1845, Lovejoy, from Massachusetts, and Pettygrove, from Maine, flipped a coin to determine if the site would be called Boston or Portland. Pettygrove won. Around this time Portland earned the nickname Stumptown, because so many stumps were left in public right-of-ways.

The California gold rush's initial impact on Portland was to nearly empty it of people, all of whom fled south seeking fortune. Pettygrove was among those who left for San Francisco in 1849, selling his stake in the city to tanner Daniel Lownsdale for a heap of leather (later sold at great markup in San Francisco). A steady stream of Oregon lumber and wheat followed the gold-seekers to San Francisco.

In 1851, Portland was incorporated as a city and soon became the dominant port of the Willamette and the Columbia Basin. Portland's population grew as white settlers, many from the Northeast of the United States, gained land under the Donation Land Claim Act of 1850. This act was an effort by the federal government to promote homesteading in the Oregon Territory by offering 320 acres at no charge to white male citizens eighteen years of age or older who occupied their claims for four consecutive years. The Donation Land Claim dispossessed Native Americans of ancestral lands and instilled a system where only white men could own property.

A plank road was partially constructed to access agricultural lands west of the city. The telegraph arrived in 1864. By the 1880s, the railroad opened markets east of the Rockies. The 1880s also brought waves of immigrants, primarily from China, Europe, and Japan.

Portland's population and land area increased dramatically in 1891 when the city consolidated with East Portland and Albina, a move partially motivated by fears that Seattle might catch up as a regional power. By 1900, Portland's population reached ninety thousand and exceeded two hundred thousand by 1910.

The map on the opposite page shows several towns just beyond the 1906 city limits that were soon annexed by Portland. These include Lents (annexed in 1912) and St. Johns (annexed in 1915). Portland would also annex Mount Tabor, a dormant cinder cone of the Boring Lava Field. Portland is one of four cities in the United States with a volcano in its city limits—joining Bend, Oregon; Jackson, Mississippi; and Honolulu, Hawaii.

Also suggested by the map, most of the wetlands south of the Columbia have been subsequently filled in. Over the river, from south to north, span the Hawthorne, Morrison, Burnside, and Steel Bridges. Visible as well are the railroad lines that enter the city's core on the east side of the Willamette.

SEATTLE

Life changed irrevocably for the Duwamish and Suquamish in 1851 when Europeans began to settle what became Seattle, a city named for Duwamish and Suquamish leader Chief Si'ahl (sometimes written as Chief Sealth). On November 13, 1851, the Denny Party landed on what is now called Alki Point. The twenty-two person group were relatives of Arthur Denny and Mary Ann Boren Denny, his wife. Traveling the rugged Oregon Trail from Cherry Grove, Illinois, to Portland, they continued on with hopes of establishing a northern outpost to connect to the westward expansion of the transcontinental railway.

By the following year most of the Denny Party had moved across Elliott Bay to establish a more sheltered site—today's Pioneer Square. Three of them, William Bell, Carson Boren, and Arthur Denny, staked claims. Most of Denny's claim became downtown Seattle. William Bell claimed an area north of downtown, called Belltown in his honor.

Early settler David S. "Doc" Maynard founded the outpost town's first hospital. In 1853, Henry Yesler, the future first mayor of the city, established a steam-powered sawmill on Mill Street, later renamed Yesler Way. The area nearby where workers dragged, or "skidded," logs down steep hills to the mill became known as Skid Road (a.k.a. skid row), the dividing line between the affluent and working-class parts of town.

The nascent town was initially called Duwamps, after the Duwamish tribe. Maynard insisted the site be renamed for Chief Seattle. Over the hilly and densely forested land, Denny, Boren, and Maynard mapped the city's first plat in 1853. Maynard's plat met the others at an angle, visible today in how streets meet at Yesler Way.

In 1861, the Territorial University of Washington was established in Capitol Hill, an area Denny had platted in hopes of making Seattle the state capital. The university moved to its present neighborhood in 1885 with the opening of Denny Hall. Four years later, it was renamed the University of Washington.

Despite the Northern Pacific Railway's 1874 decision to establish its western terminus in Tacoma, Seattle continued to grow driven by timber and coal industries and the development of independent railroads. In 1884, a spur line opened connecting Seattle to the North Pacific network, and Seattle's population continued to increase.

On June 6, 1889, a fire that began in a store on First Avenue and Madison Street quickly spread and destroyed downtown. No deaths were reported, but property damage was extensive. The city rebuilt only to struggle through the economic panic of 1893. By this time, immigrants from China and Japan had established communities in the city.

In July 1897, laden with gold from the Yukon's Klondike River, the steamship *Portland* arrived in Elliott Bay. Seattle boomed with the Yukon-Alaska gold rush, serving as the main supply center for miners and becoming the "Gateway to Alaska."

Seattle's topography was altered for urban development, through a series of regrades, which involved cutting away and leveling high points. Much of what was excavated was dumped in the bay, forming what today is the city's Central Waterfront. The massive transformation of downtown known as the Denny Regrade occurred in three phases that began in 1898 and ended in 1930.

In 1907, the city doubled in size through the annexation of Ballard, West Seattle, and Southeast Seattle. That same year a public market opened at Pike Place. In 1909, the University of Washington campus hosted the Alaska-Yukon-Pacific Exposition. By 1910, a quarter of a million residents lived in the city.

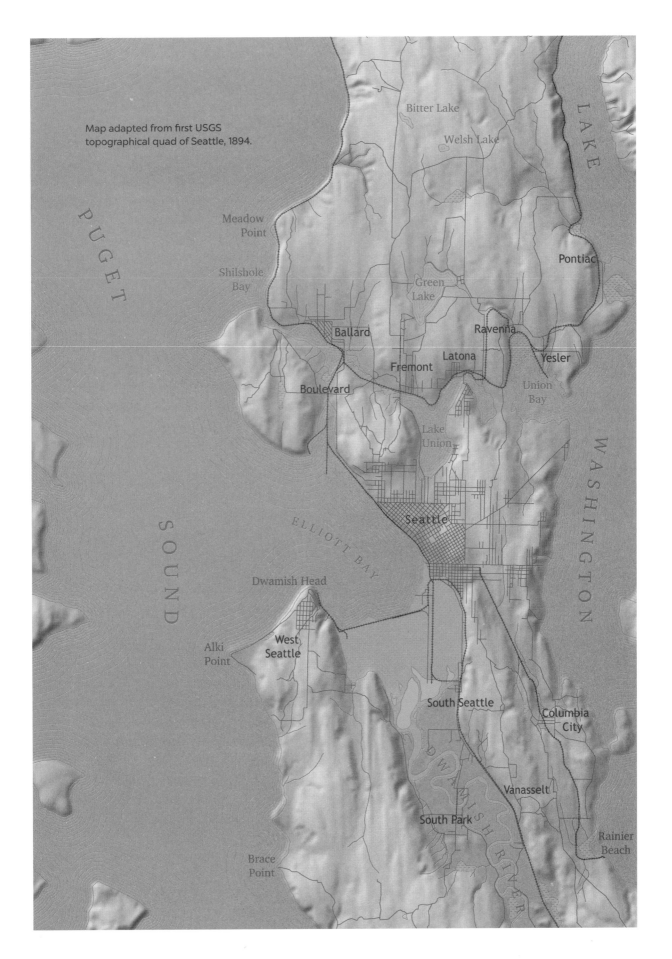

Map adapted from first USGS
topographical quad of Seattle, 1894.

PUGET

SOUND

LAKE

WASHINGTON

Bitter Lake

Welsh Lake

Meadow
Point

Shilshole
Bay

Pontiac

Green
Lake

Ballard

Ravenna

Fremont

Latona

Yesler

Boulevard

Union
Bay

Lake
Union

ELLIOTT BAY

Seattle

Dwamish Head

Alki
Point

West
Seattle

South Seattle

Columbia
City

DWAMISH RIVER

Vanasselt

South Park

Brace
Point

Rainier
Beach

WHAT IS A METRO AREA?

Straightforward question, right? A metro (metropolitan) area is the urban and suburban development around a city. How do we determine the boundaries of a metro region? That is a trickier question to answer.

Drawing these boundaries turned out to be much more difficult than we thought. Oftentimes, researchers create metro boundaries based on county boundaries because it is convenient (population statistics are often aggregated at the county level). Our collective sense of these metro regions didn't quite follow the county boundaries. So, we decided to make our own.

For each city, our final metro regions were quite disrespectful of county boundaries. How did we come up with these boundaries? One tool we used was satellite imagery, which helped us to identify major population and development clusters. We looked closely at versions of the metro regions already created. We also relied on our own knowledge of the areas surrounding the three cities.

PORTLAND AREA COUNTIES

CLARK
WASHINGTON
MULTNOMAH
CLACKAMAS

PORTLAND METRO AREA

Vancouver
Camas
Washougal
Forest Grove
Hillsboro
Portland
Troutdale
Gresham
Beaverton
Milwaukie
Happy Valley
Tigard
Lake Oswego
Gladstone
Tualatin
West Linn
Sherwood
Oregon City
Wilsonville

0   15   30   45   60 Miles

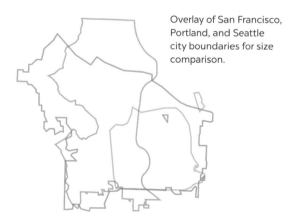

Overlay of San Francisco, Portland, and Seattle city boundaries for size comparison.

PORTLAND

Portland's urban growth boundary, a demarcation intended to limit urban sprawl, is a good guide for the Oregon side of the metro region. For this reason, Beaverton, Hillsboro, and even Forest Grove to the west all make the cut. Sherwood, Wilsonville, and Oregon City form the southern edge. To the east, we included Troutdale, Gresham, and Damascus.

We needed to figure out how much of Washington's Clark County to include. In this case satellite imagery was helpful. Vancouver is clearly part of the Portland metropolitan area, as many people commute across state lines.

SEATTLE

For Seattle's metro area, we included the Snohomish County cities of Lynnwood, Mill Creek, and Everett to the north, beyond which is urban sprawl. We left out east King County, leaving Redmond, Sammamish, and Issaquah as the eastern edge.

We decided that Bainbridge Island in Kitsap County was the westernmost extent, although we considered pushing the boundary further west. We included Tacoma, which shares both an airport, Sea-Tac, and joint seaport operations with Seattle. To the south in Pierce County we went beyond Tacoma to Lakewood after initially leaving it out of the region. We draw the southern border at the military base south of Tacoma.

SAN FRANCISCO

Our San Francisco metro area follows city boundaries more than county boundaries. There is a great deal of unincorporated space in counties such as Contra Costa and Alameda that we left out. On the south side, we included Half Moon Bay. We left off Santa Clara County south of San Jose/Los Gatos and north of Coyote/Madrone. Gilroy got chopped too. Our San Francisco region ends just before Altamont. We felt that the metro area extends north to Novato but no further.

SOMEWHERE IN THE NEIGHBORHOOD OF

In many ways, we experience cities through its neighborhoods. Some neighborhoods are widely known, even to tourists, while other neighborhoods may go unknown even to longtime residents. People who live in a city might know a few neighborhoods extremely well (where they live, work, and play) but seldom or never experience other parts of town.

San Francisco, Portland, and Seattle are known as cities with distinctive, sometimes famous, neighborhoods. Popular neighborhoods become tourist attractions. Visitors often come with lists of neighborhoods to check out and probably pick their accommodations based on neighborhood. Longtime residents often have strong neighborhood allegiances and often nostalgia for their neighborhood at a particular time.

Neighborhoods are sometimes portrayed in stark terms—as in being either good or bad. Real life is more nuanced than that. Not everyone has the same idea where a neighborhood begins and ends—or whether some neighborhoods exist at all. Official neighborhood names and boundaries are not necessarily the same ones observed by residents, and they may differ from those periodically invented to hype the next up-and-coming area (hello, East Cut).

Although drawing a map of neighborhoods would seem straightforward, that is not the case. There are official city neighborhoods but also official city subneighborhoods. There are the neighborhood maps, often with creative names, drawn up by real estate agents. There are school districts, voting districts, and zip codes. And then there are the individual mental maps of neighborhoods we each carry with us wherever we go.

The maps on the next pages tease out a view of what might be called the core or well-known neighborhoods of each city. Our goal in making these was to identify areas that are most associated with the city in the popular imagination—not to designate a "correct" way of dividing the city.

Each city has a slightly different story related to the origin, development, and current configuration of neighborhoods. For example, some neighborhoods were formerly towns of their own before being annexed; in Portland, this includes Lents, Linnton, and St. Johns; in Seattle, this includes Ballard, Ravenna, and South Park. This is not so much the case in San Francisco.

Some neighborhoods form around topographical aspects of the city. In each city we find neighborhood names related to elevation—hill, mount, peak, and valley. Other neighborhoods, including all three downtowns, are established in relation to water, while Chinatown and Japantown are examples of neighborhoods strongly associated with ethnicities. In Portland and Seattle, many neighborhoods are named for public schools.

SAN FRANCISCO

San Francisco is filled with famous neighborhoods, including the Castro, Chinatown, Haight-Ashbury, the Mission, and North Beach. Smaller neighborhoods in the southeast of the city tend to be less well known. In most cases, neighborhood borders are fuzzy; the Mission is well known, but where does it begin and end?

Neighborhoods related to elevation include Nob Hill, Potrero Hill, Russian Hill, Twin Peaks, Pacific Heights, and Bernal Heights. Homes in these areas often have expansive views of the city. Some neighborhoods have deep connections to water. The Marina District, built on landfill, used to be water. Sea Cliff affords many wealthy residents amazing views of the bay, the bridge, and the fog.

Other neighborhoods evoke a hidden past. Lone Mountain was the site of several cemeteries that few current city residents know were ever there. Today, the Western Addition encompasses a much smaller area than when it was created in the 1850s.

PORTLAND

Portland is divided into quadrants, except that there are five—no wait—six unevenly divided quadrants: NE, SE, SW, NW, North, and most recently, South Portland. This misnomer confounds visitors and residents alike. To stir the pot, we added yet another quadrant, East Portland.

The Willamette River divides the city east from west and marks a lived social division in the city. The west side has long been the wealthier side of town. Burnside divides the city north from south—marking neither natural nor social boundary. Portland's neighborhoods related to elevation include Goose Hollow, Healy Heights, Mount Scott-Arleta, Mount Tabor, and Overlook.

Some places are increasingly known by their main commercial street including Hawthorne, Mississippi, Alberta, and Division. The Portland of popular imagination sometimes stops at Eighty-Second Avenue, but Portland goes to 164th Street. East Portland past Eighty-Second Avenue is sometimes called the Numbers.

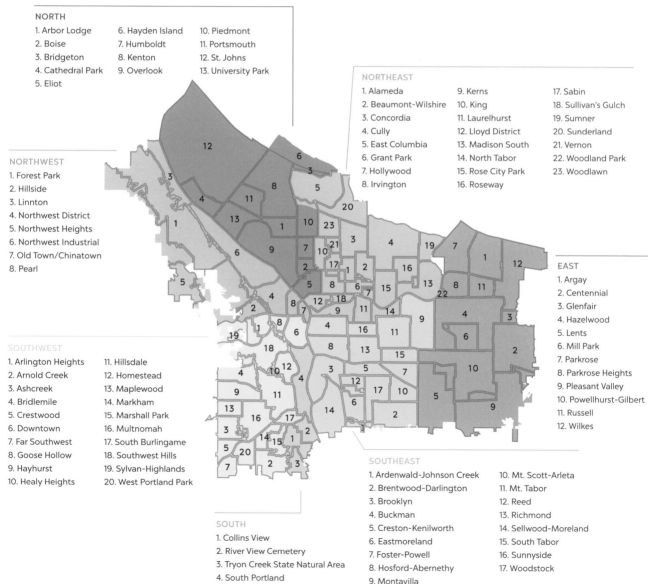

NORTH
1. Arbor Lodge
2. Boise
3. Bridgeton
4. Cathedral Park
5. Eliot
6. Hayden Island
7. Humboldt
8. Kenton
9. Overlook
10. Piedmont
11. Portsmouth
12. St. Johns
13. University Park

NORTHWEST
1. Forest Park
2. Hillside
3. Linnton
4. Northwest District
5. Northwest Heights
6. Northwest Industrial
7. Old Town/Chinatown
8. Pearl

NORTHEAST
1. Alameda
2. Beaumont-Wilshire
3. Concordia
4. Cully
5. East Columbia
6. Grant Park
7. Hollywood
8. Irvington
9. Kerns
10. King
11. Laurelhurst
12. Lloyd District
13. Madison South
14. North Tabor
15. Rose City Park
16. Roseway
17. Sabin
18. Sullivan's Gulch
19. Sumner
20. Sunderland
21. Vernon
22. Woodland Park
23. Woodlawn

EAST
1. Argay
2. Centennial
3. Glenfair
4. Hazelwood
5. Lents
6. Mill Park
7. Parkrose
8. Parkrose Heights
9. Pleasant Valley
10. Powellhurst-Gilbert
11. Russell
12. Wilkes

SOUTHWEST
1. Arlington Heights
2. Arnold Creek
3. Ashcreek
4. Bridlemile
5. Crestwood
6. Downtown
7. Far Southwest
8. Goose Hollow
9. Hayhurst
10. Healy Heights
11. Hillsdale
12. Homestead
13. Maplewood
14. Markham
15. Marshall Park
16. Multnomah
17. South Burlingame
18. Southwest Hills
19. Sylvan-Highlands
20. West Portland Park

SOUTH
1. Collins View
2. River View Cemetery
3. Tryon Creek State Natural Area
4. South Portland

SOUTHEAST
1. Ardenwald-Johnson Creek
2. Brentwood-Darlington
3. Brooklyn
4. Buckman
5. Creston-Kenilworth
6. Eastmoreland
7. Foster-Powell
8. Hosford-Abernethy
9. Montavilla
10. Mt. Scott-Arleta
11. Mt. Tabor
12. Reed
13. Richmond
14. Sellwood-Moreland
15. South Tabor
16. Sunnyside
17. Woodstock

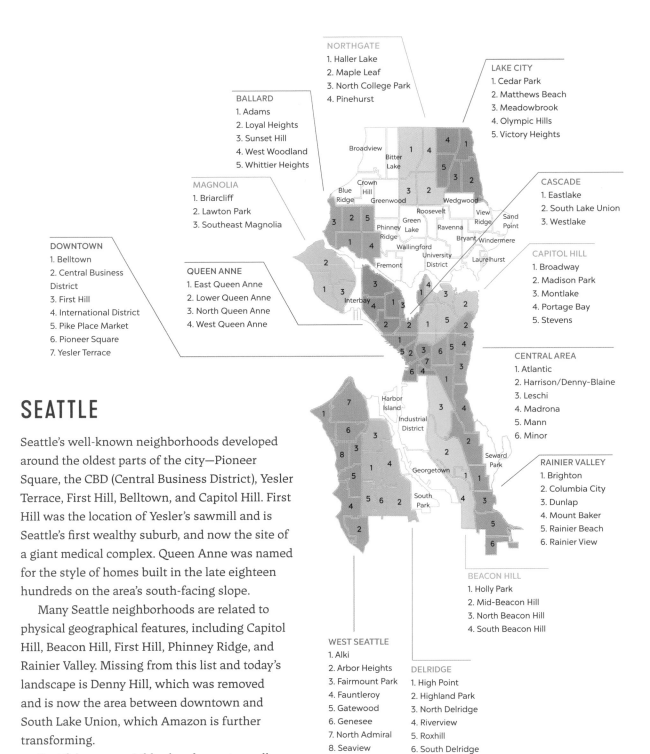

NORTHGATE
1. Haller Lake
2. Maple Leaf
3. North College Park
4. Pinehurst

LAKE CITY
1. Cedar Park
2. Matthews Beach
3. Meadowbrook
4. Olympic Hills
5. Victory Heights

BALLARD
1. Adams
2. Loyal Heights
3. Sunset Hill
4. West Woodland
5. Whittier Heights

MAGNOLIA
1. Briarcliff
2. Lawton Park
3. Southeast Magnolia

CASCADE
1. Eastlake
2. South Lake Union
3. Westlake

DOWNTOWN
1. Belltown
2. Central Business District
3. First Hill
4. International District
5. Pike Place Market
6. Pioneer Square
7. Yesler Terrace

QUEEN ANNE
1. East Queen Anne
2. Lower Queen Anne
3. North Queen Anne
4. West Queen Anne

CAPITOL HILL
1. Broadway
2. Madison Park
3. Montlake
4. Portage Bay
5. Stevens

CENTRAL AREA
1. Atlantic
2. Harrison/Denny-Blaine
3. Leschi
4. Madrona
5. Mann
6. Minor

RAINIER VALLEY
1. Brighton
2. Columbia City
3. Dunlap
4. Mount Baker
5. Rainier Beach
6. Rainier View

BEACON HILL
1. Holly Park
2. Mid-Beacon Hill
3. North Beacon Hill
4. South Beacon Hill

WEST SEATTLE
1. Alki
2. Arbor Heights
3. Fairmount Park
4. Fauntleroy
5. Gatewood
6. Genesee
7. North Admiral
8. Seaview

DELRIDGE
1. High Point
2. Highland Park
3. North Delridge
4. Riverview
5. Roxhill
6. South Delridge

SEATTLE

Seattle's well-known neighborhoods developed around the oldest parts of the city—Pioneer Square, the CBD (Central Business District), Yesler Terrace, First Hill, Belltown, and Capitol Hill. First Hill was the location of Yesler's sawmill and is Seattle's first wealthy suburb, and now the site of a giant medical complex. Queen Anne was named for the style of homes built in the late eighteen hundreds on the area's south-facing slope.

Many Seattle neighborhoods are related to physical geographical features, including Capitol Hill, Beacon Hill, First Hill, Phinney Ridge, and Rainier Valley. Missing from this list and today's landscape is Denny Hill, which was removed and is now the area between downtown and South Lake Union, which Amazon is further transforming.

Seattle's macroneighborhoods are generally better known than the subneighborhoods that compose them. For example, everyone knows West Seattle, but which eight subneighborhoods compose it? Even downtown Seattle has seven subneighborhoods.

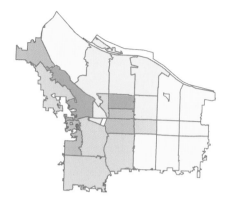

Median house price, January 2019

$2,515,800
$1,233,100
$789,600
$485,600
$300,800

PROFILES IN DEMOGRAPHY

Houses in San Francisco, Portland, and Seattle are expensive. In January 2019 the median house price in the United States was $229,000. In Portland, the median house price was $416,000. In Seattle it was $714,200. That figure in San Francisco was (wait for it) $1,351,900. In some areas beyond city limits we would find even higher median house prices.

We normalized the median house price scale across the three cities to make comparing across cities easier. The dark blue color is the higher end of the price range with a median house price of more than $2.5 million. The light yellow and green colors are at the lower end of the price range at $300,000 (less expensive maybe, but well above the national average).

The areas with the lowest median house prices in San Francisco far exceed the highest median house prices in Portland. Seattle is somewhere in between—the median house price for all three cities is $714,100, nearly identical to Seattle's own median price.

It doesn't take looking at this for long to understand why so many people from San Francisco have decided, after years of making San Francisco money but paying San Francisco prices, to relocate to Portland or Seattle. Or to understand the alarm of people in Portland and Seattle, who recognize that an $800,000 house is a bargain to people in San Francisco. The tension is real.

Portland's median household annual income is $900 less than the national median of $60,336. The median household annual income of Seattle is just over $79,000. In San Francisco that figure is $98,000.

All three cities far exceed the 9 percent national average of carless commuting. In Portland, the figure is 20 percent and in Seattle 30 percent. Meanwhile in compact San Francisco where parking is brutal in every neighborhood, half of commuters don't use cars.

Portland overall has 38 percent renters, quite close to the national average of 36 percent. By comparison, Seattle has about 47 percent renters and San Francisco has about 63 percent renters.

PORTLAND

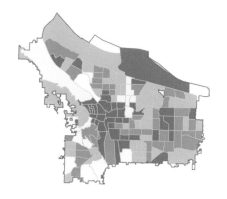
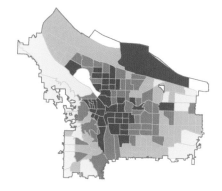
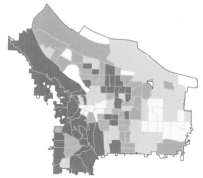

SEATTLE

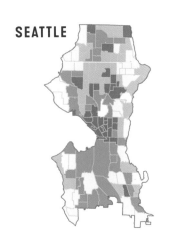
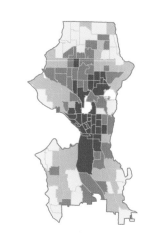
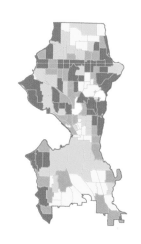

SAN FRANCISCO

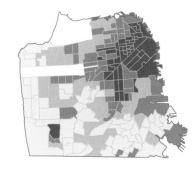
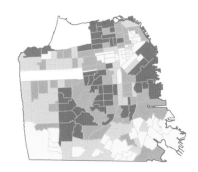

Renters (in percent)

	PDX	SEA	SF
	8–20	6–30	5–42
	20–38	30–47	42–63
	38–53	47–64	63–80
	53–90	64–100	80–100

Carless commuters (in percent)

	PDX	SEA	SF
	3–11	7–24	19–38
	11–20	24–30	38–50
	20–31	30–43	50–66
	31–78	43–76	66–91

Median household income (in dollars)

	PDX	SEA	SF
	13k–43k	11k–62k	12k–72k
	43k–59k	62k–79k	72k–98k
	59k–78k	79k–102k	98k–129k
	78k–171k	102–168k	129k–195k

STATISTICAL MEDLEY

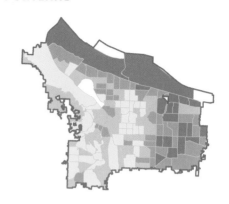

A common diversity statistic computed by the US Census Bureau is the percentage nonwhite, which we map here. The national figure for the nonwhite population is 39 percent. Compared to the national statistic, Portland's and Seattle's percentages are much lower at 23 percent and 26 percent respectively. This positions Portland as one of the whitest cities in the country. San Francisco, by contrast, is 51 percent nonwhite.

In San Francisco, Portland, and Seattle, there are a bunch of clichés that center on baristas having master's degrees and PhDs. They aren't meant to be funny. They are meant to scare college graduates away (which does not seem to be working). We don't have statistics on baristas with graduate degrees, but we can report that nationally the percentage of people with bachelor's degrees or higher is 32 percent. Portland's own percentage is 49 percent, San Francisco's is about 58 percent, and Seattle's is 60 percent.

The number of children living in each city has declined over the past several decades. Contrary to what you might have heard, there are still children living in San Francisco. However, the median age there is 39.1 years (almost 40!) whereas the median age in the United States is 38.1. Portland is a little younger with a median age of 36.2, as is Seattle with a median age of 37.

In San Francisco, Portland, and Seattle, locals of each city will often tell you that no one is from there anymore. Although not technically correct, statistics do back the claim that people from out of state are in the majority. Nationally, the percentage of people who were born in the state where they currently live is 58.5 percent; that number is much lower in Portland (37.5 percent), Seattle (38 percent), and San Francisco (39 percent).

SEATTLE

SAN FRANCISCO

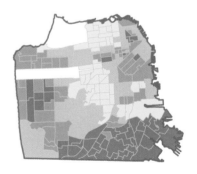

Diversity (in percent)

	PDX	SEA	SF
	34–53	44–91	67–96
	23–34	26–44	51–67
	14–23	18–26	33–51
	5–14	8–18	11–33

	Bachelor's degree-holders or higher (in percent)			Resident median age (in years)			Residents born in-state (in percent)		
	PDX	SEA	SF	PDX	SEA	SF	PDX	SEA	SF
	10–21	12–28	9–41	22–33	20–34	20–36	16–30	15–34	11–34
	21–50	28–60	41–57	33–36	34–37	36–39	30–38	34–38	34–39
	50–64	60–73	57–72	36–40	37–41	39–43	38–44	38–44	39–44
	64–85	73–87	72–90	40–54	41–54	43–66	44–54	44–55	44–74

GLOBAL POSITIONING

Let's consider how Upper Left cities are positioned relative to each other and to other places in the world. San Francisco is 535 miles as the crow flies (or 635 miles if you drive) from Portland, which is 145 miles (175 in a car) from Seattle.

Portland (W 122°41'), San Francisco (W 122°25') and Seattle (W 122°20') are within less than half a degree of longitude of one another. Across the globe the only other big city in that half a degree of longitude is Oakland.

But what other cities are within half a degree of latitude of our three? The map below shows a selection of these cities. How places line up might surprise you.

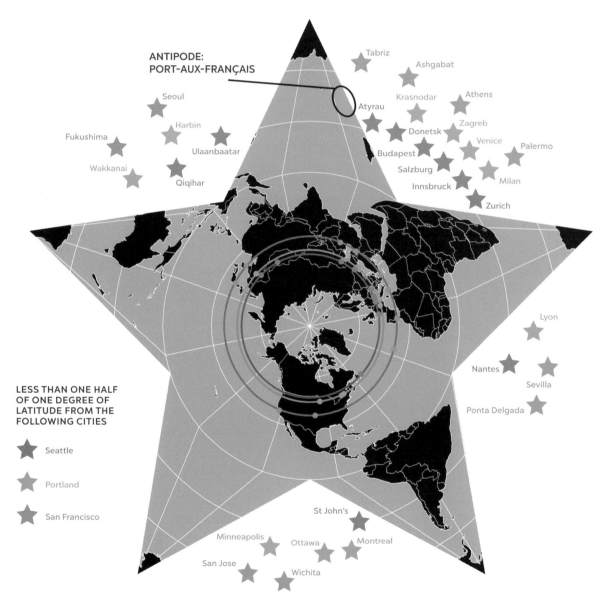

ANTIPODE:
PORT-AUX-FRANÇAIS

Tabriz · Ashgabat · Krasnodar · Athens · Atyrau · Zagreb · Donetsk · Venice · Palermo · Budapest · Salzburg · Milan · Innsbruck · Zurich

Seoul · Harbin · Fukushima · Ulaanbaatar · Wakkanai · Qiqihar

Lyon · Nantes · Sevilla · Ponta Delgada

LESS THAN ONE HALF
OF ONE DEGREE OF
LATITUDE FROM THE
FOLLOWING CITIES

⭐ Seattle

⭐ Portland

⭐ San Francisco

St John's · Minneapolis · Ottawa · Montreal · San Jose · Wichita

Port-aux-Français, Kerguelen Islands, 2018. The farthest permanent settlement in the world from San Francisco, Portland, and Seattle.

ANTIPODES: WANNA GET AWAY?

Antipodes are opposite points on a globe. The antipodes of San Francisco, Portland, and Seattle are in the frigid Indian Ocean waters, just outside the Antarctic Circle.

But what if you wanted to get as far away from the Upper Left as possible and still be on land? The closest settlement to San Francisco, Portland, and Seattle's antipodes is Port-aux-Français of the Kerguelen Islands (a.k.a. the Desolation Islands).

Among the most isolated places on Earth, the French territory is nearly equidistant from Africa, Antarctica, and Australia and only accessible by boat. The next closest settlement is over 2,000 miles away in Madagascar.

Pack for short, cloudy, and cold summers followed by long, somewhat cloudy, and cold winters. Expect wicked winds year-round.

Rampant populations of elephant seals, royal penguins, and rabbits mingle with 45 (in winter) to 120 (in summer) humans. A permanent weather base and thousands of sheep also inhabit the island. Hope you like mutton.

I. URBAN LANDSCAPES

MANY PEOPLE MAY THINK OF *landscape* as a term that relates to pastoral paintings or wild vistas, but cities have landscapes too—what geographers call urban landscapes. Embedded in these urban landscapes are stories of loss, self-expression, and belonging.

Urban landscapes reveal more than preferences for building styles—they tell us something about history, ideologies, and power. Of course, there isn't just one history, one ideology, one story, but many.

Studying urban landscapes over time reveals fascinating patterns of change in a city. Although urban landscapes are always in flux, changes are sometimes quick and dramatic. Many areas of San Francisco, Portland, and Seattle feel transformed overnight; a few areas never seem to change. By paying close attention to the details of everyday city life, we can better understand the mosaic of urban landscapes.

In this chapter, we explore urban landscapes in San Francisco, Portland, and Seattle by exploring topics that range from city block shapes to public spaces to cemeteries.

BREAKING THE GRID

The graphics in this section were inspired by the French artist Armelle Caron, who takes a map of city blocks and disassembles them into many puzzle-like pieces. Caron's arrangements, what she calls graphic anagrams, provide a nonconventional way of looking at a city. We employed this technique in San Francisco, Portland, and Seattle. We took the downtown of each city and rearranged the blocks. We did the same with transects of Market Street, Sandy Boulevard, and Madison Street to illustrate the blocks along a diagonal street in each city.

Madison Street, Seattle

SEATTLE

Seattle's downtown borders the waterfront, which creates some of the very large shapes that contrast strongly with the grids of small rectangles that make up most of downtown. The long shape with a triangular end contains Elliott Bay Park and Myrtle Edwards Park, which run northwest from downtown. The shifting grid of Seattle's downtown makes for some interesting shapes. Madison Street borders a number of large properties, of which the largest and most oddly shaped is a golf course and neighboring arboretum. Piers from Lake Washington are also evident on the next largest shape.

Downtown Seattle

SEATTLE'S CURIOUS URBAN FORMS

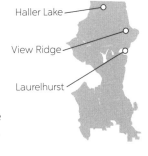

LAURELHURST

Laurelhurst is located on a small peninsula on the shore of Lake Washington, east of the University of Washington. Real estate developers coined the name of the area in 1906. King County's first Sheriff, William H. "Uncle Joe" Surber, lived on the land in the 1860s, where he hunted deer and bear. In 1888, Henry Yesler established the Town of Yesler on the site and established a sawmill that cleared the area of trees. The McLaughlin Realty Company platted the area and named it. Housing development started in 1906 and the area was annexed by Seattle in 1910.

VIEW RIDGE

View Ridge was platted and developed in the mid-1930s by a couple of radio station employees who decided to get into real estate. The partners cleared the land of trees, exposing amazing vistas of Lake Washington and the Cascades. To preserve views, houses were set back from the street. Seattle annexed the area in 1942. Evidence of at least one covenant dated 1950 from the neighborhood has surfaced with the language, "No person of any race other than the white Caucasian race shall use or occupy any building or lot." The oval streets are now almost completely surrounded by a country club.

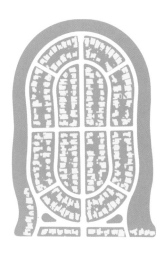

HALLER LAKE

Haller Lake is both the name of a small neighborhood and a smaller lake in North Seattle. Lot sizes are large, corresponding with the farming history of the area. In 1905, Theodore N. Haller bought and platted the area, including a kind of sloppy hexagon around the lake. It took decades for residents to fill the area. The relatively undeveloped area held limited appeal for people in the city. A small park connects Haller Lake to the city grid at N 125th Street. Seattle annexed the area in 1954, four years after the opening of the nearby Northgate Mall.

SAN FRANCISCO

Downtown San Francisco's myriad rectangular blocks contrast with the long shape along the Embarcadero Freeway that includes the detailed outlines of piers jutting into the bay. The rectangular blocks spread out over a number of large hills, yet the grid remains intact. Market Street borders some of the large rectangular blocks of downtown as well as a number of blocks with visible alleys. As Market Street leaves downtown and runs up Twin Peaks, the block shapes it encounters start to get curvy.

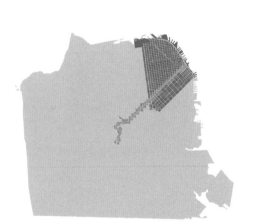

Market Street, San Francisco

Downtown San Francisco

SAN FRANCISCO'S CURIOUS URBAN FORMS

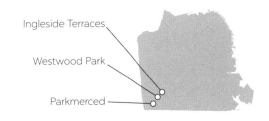

PARKMERCED

Parkmerced is a planned community in San Francisco designed in the 1940s as a car-based suburbia within the city limits. Framed by Lake Merced Park, San Francisco State University, Nineteenth Avenue, and the San Francisco Golf Club, the development includes a series of high-rise apartment buildings built to accommodate demand post World War II. At the core of the development is Juan Bautista Circle. From the 1960s through the 1990s, the area suffered from its car-centric design and the neighborhood became neglected. More recently the area has begun to redevelop with plans to restore the neighboring watershed as well.

WESTWOOD PARK

In 1916, building began on a residential development called Westwood Park. It is located just northeast of where Parkmerced was later developed. The opening of West Portal's Twin Peak tunnel in 1918 made the area accessible to the neighborhoods further north. As early as 1920, racially restrictive covenants prohibited nonwhite people from buying property or renting in the neighborhood. More than 650 bungalows line the streets of the neighborhood where it is still prohibited to construct apartment buildings. The two oval streets at the center of the development were added to avoid a tangle of confusing streets.

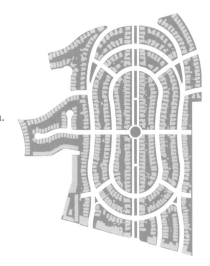

INGLESIDE TERRACES

Ingleside Terraces, located between Parkmerced and Westwood Park, is a neighborhood of single-family homes developed beginning in 1910 on the site of the Ingleside horse track (also used for car racing). The last horse race occurred on December 30, 1905, and the following year the track served as a refugee camp for victims of the 1906 earthquake and fire. Urbano Drive was paved in the space of the track and maintains the shape. The west side of the oval contains a park with a large sundial. The neighborhood had racially restrictive covenants barring nonwhites from buying or living in the neighborhood.

PORTLAND

Downtown Portland has smaller blocks than the downtown areas in San Francisco or Seattle. This creates an assemblage made mostly of squares rather than rectangles. Portland was platted with unconventionally short blocks because smaller blocks meant more corners and thus more corner properties, which fetch higher rents. Larger blocks are on the downtown's periphery. The longest shape borders the Willamette River. The transect of Sandy Boulevard terminates near an outdoor Catholic sanctuary and shrine, the Grotto, and the adjacent Rocky Butte Natural Area, which is the large shape in the center of the graphic.

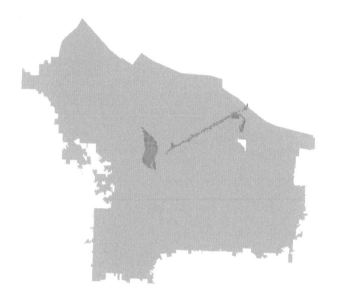

Sandy Boulevard, Portland

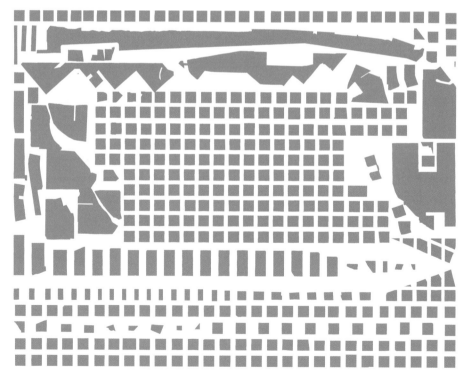

Downtown Portland

PORTLAND'S CURIOUS URBAN FORMS

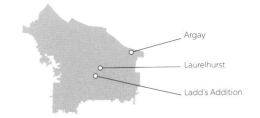

Argay
Laurelhurst
Ladd's Addition

LADD'S ADDITION

Ladd's Addition is a streetcar neighborhood in Southeast Portland platted in 1891. The developer and landowner, William S. Ladd, sought to transform his farm into an exclusive neighborhood for the wealthy. Contrary to his surveyor's advice, Ladd opted for blocks that created an X with two primary diagonal streets. The design included five park areas and servant alleys between blocks to help promote the neighborhood as elite. The first houses were built in 1905, twelve years after Ladd's death. Early prohibitions restricted people of Chinese and Japanese ancestry from living in the neighborhoods; these restrictions expired in the 1930s.

LAURELHURST

Laurelhurst is another streetcar suburb developed on Ladd's farmland. Ladd established a 464-acre dairy farm on the site, which his son sold in 1909. The neighborhood design was informed by ideas provided by landscape architect John Charles Olmsted. The largest open area on the map is Laurelhurst Park, designed in 1912. Large sandstone pillars still mark the western and southern entrances to the neighborhood. A gilded statue of Joan of Arc, a memorial to World War I, sits in a traffic circle. Like Ladd's Addition, restrictions prohibited nonwhites from living in this neighborhood for decades.

ARGAY

Argay is a somewhat obscure neighborhood in Northeast Portland that borders the Columbia River. The first names of developers Art Simonson and Gerhardt (Gay) Stabney were mashed together to create the neighborhood name Argay. The area was planned as a suburban alternative to the hectic city grid. Designed in the 1950s, it features short, curvy streets and even has cul-de-sacs. The location affords excellent views of the Columbia River, Mount Hood, and Mount St. Helens. Much of the development is midcentury single-family homes, although apartment buildings and condos pepper the neighborhood as well.

WATER UNDER THE BRIDGE

Bridges connecting points of land separated by various bodies of water figure prominently on the landscapes of San Francisco, Portland, and Seattle. In the Bay Area, some pretty big bridges are needed to do that. Portland bridges span rivers and aren't as long. Seattle's bridges span a lake, the lake's ship canal, and a river, so the sizes and lengths of bridges vary more.

A couple of Seattle's bridges float. The longest floating bridge in the world is Seattle's Evergreen Point Bridge—officially the Governor Albert D. Rosellini Bridge, but often just referred to as the 520 Bridge. Spanning Lake Washington, the bridge is supported by enormous watertight concrete pontoons secured by anchors. The world's second-longest floating bridge, the close-by Lacey V. Murrow Memorial Bridge, is part of I-90 and also crosses the lake. Conventional suspension bridges wouldn't be practical in these locations,

so the pontoon approach was adopted. Seattle also boasts the world's first hydraulically operated double-leaf concrete swing bridge—the Spokane Street Bridge.

Portland has many bridges close together, and a wide variety of bridge types, including bascule, cantilever, deck truss, tied arch, suspension, and vertical-lift. The Steel Bridge is the country's oldest and only operating telescoping vertical-lift bridge—that means the lower deck, for trains and pedestrians, can be raised without lifting the upper deck, used by light-rail, buses, and cars. The newest bridge in the city is Tilikum Crossing, Bridge of the People; it is the longest car-free bridge in the United States. The city's most beautiful bridge, the St. Johns Bridge, lies in the northern part of the city.

(right) The inset maps display the locations of each bridge and their heights above water.

(bottom) These graphics display the relative and cumulative lengths of the water-crossing bridges of San Francisco, Portland, and Seattle, as well as their heights above water.

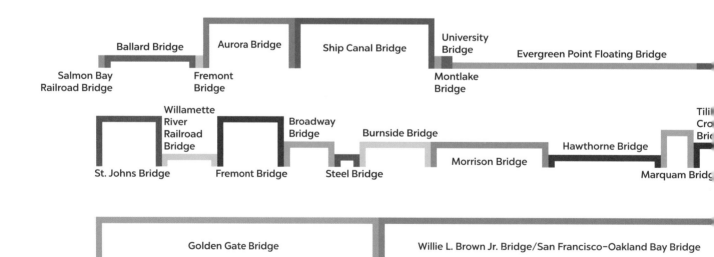

The Golden Gate Bridge is one of the most recognizable bridges in the world and, although barely in the city proper, is probably San Francisco's best-known work of architecture. Originally, the Navy wanted the bridge painted with black and yellow stripes to heighten visibility for ships; however, the vermilion hue called international orange prevailed. The San Francisco–Oakland Bay Bridge, widely known as the Bay Bridge, has two sections that meet on Yerba Buena Island. In 2013, the western half was named for former San Francisco mayor Willie L. Brown Jr.

SEATTLE'S BRIDGES

SAN FRANCISCO'S BRIDGES

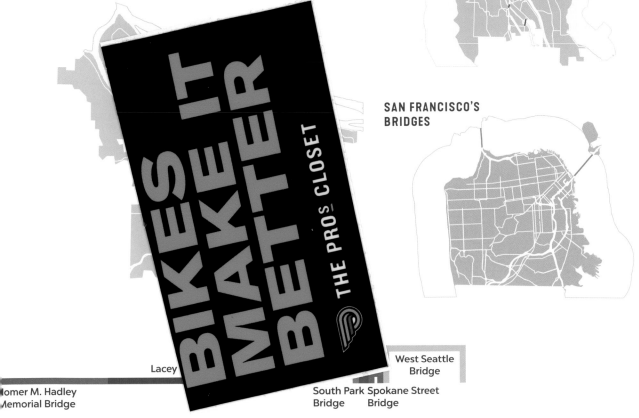

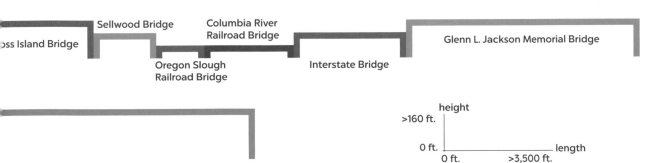

Lacey

West Seattle Bridge

Homer M. Hadley Memorial Bridge

South Park Bridge

Spokane Street Bridge

Ross Island Bridge

Sellwood Bridge

Columbia River Railroad Bridge

Oregon Slough Railroad Bridge

Interstate Bridge

Glenn L. Jackson Memorial Bridge

height
>160 ft.

0 ft.

0 ft.   >3,500 ft.

length

TRAFFIC LIGHT TAPESTRY

These colorful illustrations may look like random kaleidoscopes of color, but the shapes in these maps represent the proximity of traffic lights to one another throughout the city. Each polygon corresponds to one traffic signal, and the size of the polygon represents the distance to the next traffic signal.

If a polygon is large, it means that the traffic signal it represents is far away from other traffic signals. If the polygon is small, it means there are many other traffic signals nearby, and so the area that the traffic signal serves is much smaller, as it bumps up against the areas served by other traffic signals.

Looking at the weave of the tapestry, we can start to make out the form of the city. Downtown cores are easy to identify by their dense weave of uniform small polygons, representing the many traffic lights crowding the urban center. By contrast, we can also identify more rural areas, with sprawling polygons representing large spaces between traffic lights. Large natural areas, such as Forest Park in Portland and the Presidio in San Francisco, might have few or no traffic signals and show up as large polygons.

As the insets on this page illustrate, the shapes of streets also emerge. In San Francisco, the grid of downtown is evident just north of Market Street, which cuts a diagonal swath across the tapestry. The wide-open spaces and relative dearth of traffic lights of Southwest Portland stand out clearly. We can also pick out Seattle's Rainier Avenue, which runs north to south along the southeast part of town.

SAN FRANCISCO: DOWNTOWN NORTH OF MARKET STREET

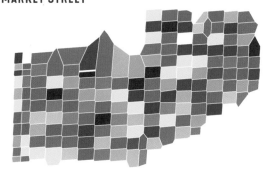

PORTLAND: SOUTHWEST QUADRANT

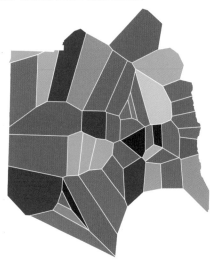

SEATTLE: RAINIER AVENUE

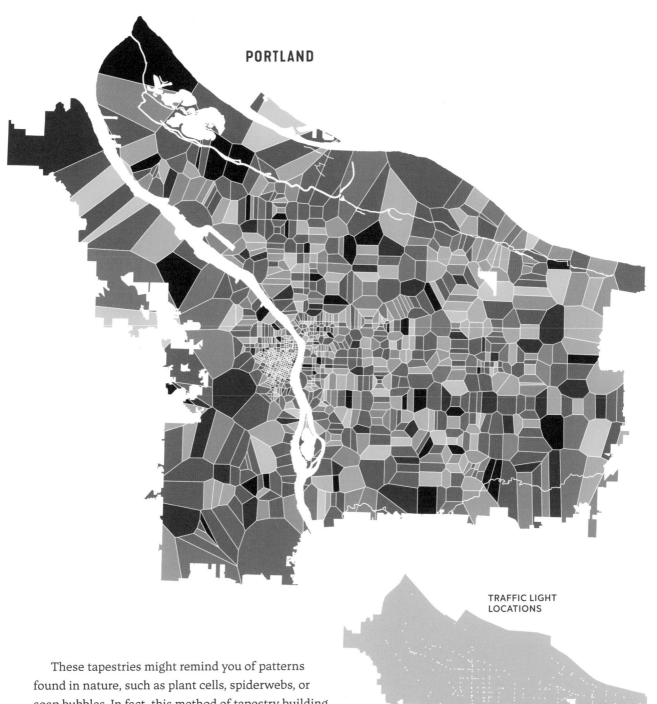

PORTLAND

TRAFFIC LIGHT LOCATIONS

These tapestries might remind you of patterns found in nature, such as plant cells, spiderwebs, or soap bubbles. In fact, this method of tapestry building (a tool known as Thiessen polygons) is used in the sciences to measure catchment areas or visualize the distance between sample points of things such as soil type or air quality. Applying this method to city infrastructure, such as traffic lights, allows us a different way of conceptualizing urban form with a psychedelic twist on density mapping.

39

SEATTLE

TRAFFIC LIGHT
LOCATIONS

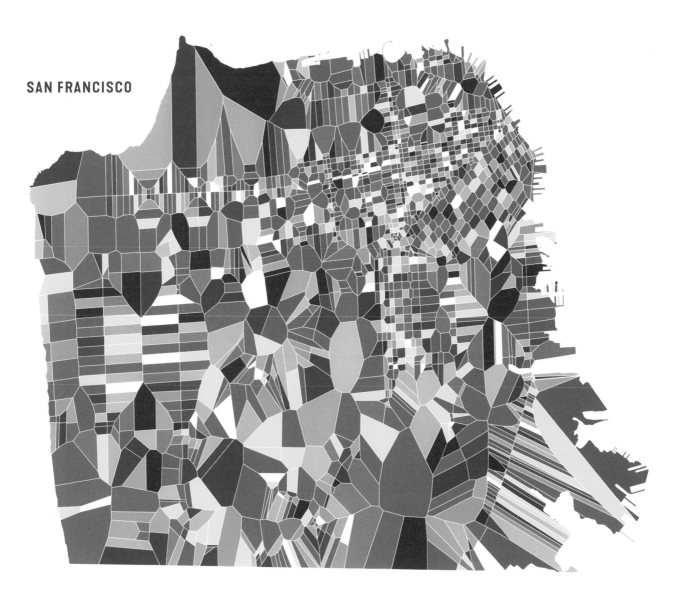

SAN FRANCISCO

TRAFFIC LIGHT
LOCATIONS

NO WASTE OF SPACE: ALLEYS

Alleys, the smallest unit of the city's transportation network, cut through and between spaces where roads do not. The change in scale from the cacophony of a busy street to the muted quiet of a back alley is a visceral transition to something that is not quite street and not quite structure. As much for wholesome communal gathering as for illicit activities, they are a liminal urban space, a threshold between public and private, between motion and stillness.

These hidden places have at times provided a meeting place for marginalized groups, a canvas for unsanctioned art, and a home for unseemly behavior. In popular culture, alleys are places to sell drugs, get in a knife fight, or dump a body. In fact, alleys are, and always have been, home to thriving microcommunities.

In the nineteenth century, alleys were built to hide the undesirable elements of urban life: collecting waste and making deliveries. In residential areas, alleys were used as transportation access lanes, as servant entrances, and as areas for children to play. This began to change with the rise of the automobile. As highways expanded and people began to move from the city to the suburbs, alleys became forgotten, underused spaces.

Portland, Seattle, and San Francisco—like many US cities—are reclaiming their alleys, turning them into centers for retail and converting them into "green alleys" by removing pavement, introducing native plants, and creating a venue for murals, wheatpasting, and stenciling. The intimacy of alleys often makes them ideal gathering spaces, and the cultural shift toward valuing the pocket-size has reinvigorated them.

It is not so much that the alley itself has changed over the years, but rather our use of it and our sense of its purpose that shifted. The alley invites exploration, suggests a change in perspective, and provides an alternate path as you move through the urban environment.

SAN FRANCISCO

San Francisco has many types of alleys—some built as utilitarian lanes for commercial and retail deliveries, some used as shortcuts between long city blocks, others as stairways carved into the steep terrain. In hilly San Francisco, alleys often take the form of steps; the city has more than four hundred public stairways, many of them hidden.

The Filbert Steps, which ascend Telegraph Hill to Coit Tower, are one of the most celebrated stairways in the city. Bordered by greenery and lined by backyard gardens, the path of the Filbert Steps is lush and colorful. Several streets in the city turn into stairways and the houses along them are accessible only from those steps.

Chinatown is home to a maze of forty-one alleys. Ross Alley, the city's oldest alley and a popular tourist destination, was infamous in the Barbary Coast days as a hot spot for gambling and brothels. Today Chinatown's alleys are used as gathering spots for residents, where laundry hangs to dry, and local workers take their breaks.

Waverly Place, also known as the Street of Painted Balconies, is an alley that was featured in the 1989 book *The Joy Luck Club*, which was later made into a movie. Jack Kerouac Alley was historically used for garbage dumping, until it was renovated in 2007 as a pedestrian walkway. The alley is now known for its decorative lampposts and its inscriptions (written in both English and Chinese) by famous writers.

Other alleys of San Francisco have made a name for themselves as public art galleries. Veterans Alley, a community-based mural project located in San Francisco's Tenderloin District, is a place where veterans from around the world tell their stories through art.

Clarion Alley and Balmy Alley are located in the Mission District, which has the most concentrated collection of murals in the city. In the early 1970s, a small group of Latina muralists painted Balmy Alley, which continues today and features murals related to Central American political struggles. Ranging in styles and subjects, Clarion Alley has increasingly become dedicated to themes of social justice, human rights, and critique of local gentrification. As such, these alleys have become sites of resistance, where the politics of the neighborhood are writ large on the walls.

Alley, Sunset District, San Francisco, 2017.

— Alleys
- Stairs

SEATTLE

Seattle's numerous alleys were originally conceived of as lanes for commercial use—garbage service, deliveries, and the like. Seattle is also a city of more than 650 staircases. Over the years, many alleys became neglected. This disrepair is due in large part to the fact that alleys typically do not get the funding that city streets get. In Seattle, there is no funding for repairs of paved alleys beyond spot repair (filling potholes), and unpaved alleys receive no city funding at all.

Local advocates for alleys believe that Seattle's alleys can be more than purely utilitarian. Recent alley "activation" efforts seek to rehabilitate the city's most beloved alleys. Improvement projects are underway in Chinatown, the University District, and Pioneer Square, with dedicated funding from the city government and community partners.

Seattle's most famous alley is surely Post Alley, a prominent feature of Pike Place Market in downtown Seattle. Named after the *Seattle Post*, which was once located at its southern end, Post Alley historically had a bad reputation, prompting one *Seattle Times* columnist to write in 1978, "there is not a scuzzier stretch of lowdown street in the city."

In 1979, the alley was renamed and rebranded in an effort to attract businesses to open storefronts. Post Alley is now a destination for visitors and home to Seattle's Market Theater Gum Wall, where people go to stare at, and contribute to, a growing collection of chewing gum wads several inches thick.

Historically, Canton Alley was a center of commercial, residential, and community-based activities in Chinatown, where children played and people sold produce. Over the years, the alley became riddled with potholes and choked with garbage bins, and businesses with storefronts on the alley slowly moved away. More recently, community stakeholders and business owners have helped rally funding to repave Canton, Nord, and Pioneer Passage Alleys.

Canton and Nord have been designated as "festival streets" by the Seattle Department of Transportation, so are permitted to hold pedestrian-friendly events and festivals. This is all part of a larger effort to change the perception of alleys and the culture around them—from dark, grimy, hidden, and sometimes subversive places to whimsical public spaces that encourage business and development, spaces that would be described as "vibrant" or "lively" on Yelp.

(left) Upper Post Alley (behind Stewart House looking north), Seattle, 1978.

(right) Upper Post Alley (behind Stewart House looking north), Seattle, 2019.

Alleys
Stairs

Alley, Ladd's Addition, Portland, 2019.

Alley, Concordia, Portland, 2019.

PORTLAND

Portland is not known as an alley town. Unlike Seattle and San Francisco, it has few, if any, famous alleys and almost no alleys downtown. The majority of Portland's alleys are in residential areas, several miles from the inner core of the city. Few alleys exist in older, more established central parts of the city. However, the alleys that do exist in these areas are often paved and maintained, unlike alleys in the outer neighborhoods.

The most notable cluster of alleys in the inner east side is those of Ladd's Addition, which are better maintained than some streets in other parts of the city. The neighborhood is Portland's oldest planned residential development, built between 1905 and 1930 during Portland's streetcar era. It was envisioned as "a residential section for cultured people," and its blocks were split by service alleys, intended to add to the perception of Ladd's Addition as a high-class neighborhood. Garages in Ladd's Addition were almost always built in the alley, many of them small-scale renditions of the home they accompanied. The alleys were likely used as entrances for carriages (later automobiles) and servants, maids, cooks, and other workers at the homes, so that they would not be seen coming and going from the front of the home.

There are significant concentrations of blocks with alleys located in Northeast and outer Southeast Portland in residential neighborhoods. These alleys, most likely built for garbage service and access to rear garages, usually bisect residential streets midblock and are often narrow and unpaved. They are imprints of an earlier era of urban development. Many of these residential alleys have become overgrown and unused, though they can serve as an alternative "secret" connection for pedestrian travelers.

Recent years have brought new interest in revitalizing these hidden paths. The Portland Alley Project is one notable group, formed in 2013, who support a variety of grassroots, neighborhood-scale alley revitalization efforts. In Portland fashion, alley activists keep it weird with events such as the Alleyways Goat Work Party, where a herd of goats teamed up with neighbors to clear an alley of unwanted vegetation.

Alleys

Stairs

PUBLIC SPACES: THE COMMONS, THE WATERFRONT, AND THE EMBARCADERO

Underneath any urban history are plans that could have been, once were, and never should have been in the first place. Scratch the surface of your favorite city and you will find a rich history of destruction, development, and imagined futures made possible by the creativity and determination of planners, policy makers, and the grassroots advocacy of local experts: the residents.

Public spaces have many functions, but above all they are spaces that people are able to occupy without paying anything. The famed public spaces of ancient Greek cities were the agoras, which served as marketplaces and areas to assemble. In ancient Roman cities it was the forums—places to go and openly exchange ideas. Throughout the United States, the agora and the forum both evolved into plazas, parks, and sidewalks.

These following tales suggest that the public spaces we experience in Upper Left cities today were not built overnight. And they may spring up in areas formerly used for very different functions. Attitudes toward these spaces are enmeshed in politics, economic pressures, and the realities facing haves and have-nots.

THE TRAGEDY OF THE COMMONS

Seattle has been trying to solve the "Mercer Mess" since the 1960s. This is the Mercer Corridor where I-5 meets Seattle Center. An improvement project concluded in 2015, but few people believe that has helped all that much, because the area is still a mess. This is the same area that Amazon has been developing in recent years.

Amazon's tech hub in Seattle, known as South Lake Union (SLU), was almost something else entirely. Back before the sterile streets and towers of reflective glass, the high-priced condos, and the spendy cuisine, the city considered an alternative future for the area. The Seattle Commons plan, a proposal to connect South Lake Union to downtown, outlined a sixty-one-acre greensward.

The park itself was designed with meandering pathways, tree-lined streets, an open meadow, playgrounds, and a formal garden. New businesses and jobs were promised to revitalize the surrounding areas, with the expectation of attracting new residents. The Seattle Commons plan was a New Urbanist dream that suggested residents would have no need for a car, as they could live, work, and play in the same neighborhood. Residents voted against a levy to fund the plan, stunting the utopian dream in the mid-1990s. Looking back at the plan, some residents regret that the park never came to fruition.

When the plan was introduced, SLU voters weren't against the park itself. Rather, they were cautious to fund a scheme designed to benefit developers. Vulcan Inc., owned by Microsoft billionaire Paul Allen, took ownership of a large section of land in the proposed project as a pledge of buying in to the proposal. To acquire the land, Allen offered a loan of $25 million for the project with the stipulation that the funds would be forgiven should the vote pass. Another $111 million was needed to complete the work. For voters at the time, shutting down this project was a win for small businesses. Upon rejection, Vulcan maintained ownership of the land and continued to acquire more land in SLU, and has since developed condos and office buildings supporting the Amazon campus.

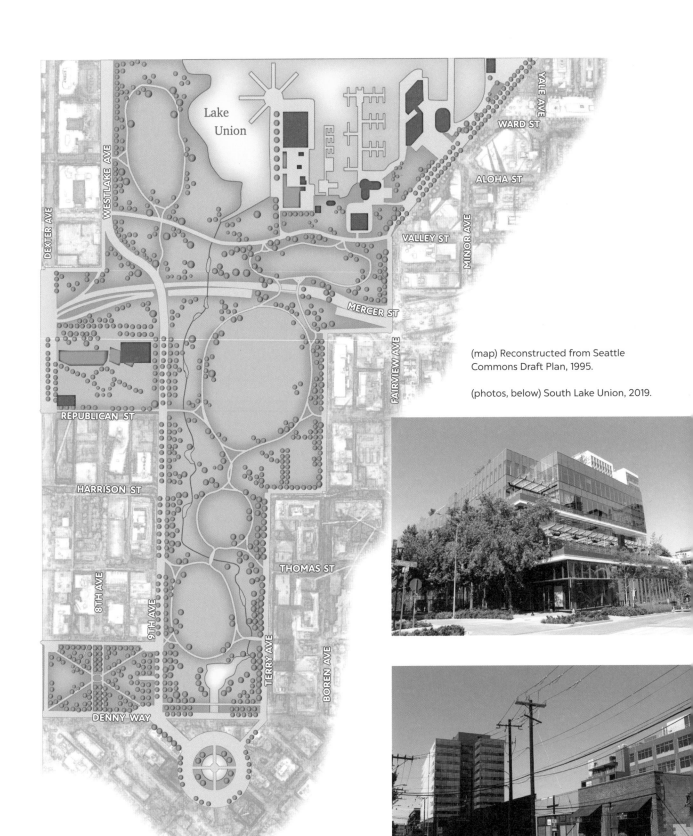

Lake
Union

YALE AVE

WARD ST

ALOHA ST

WESTLAKE AVE

DEXTER AVE

MINOR AVE

VALLEY ST

MERCER ST

FAIRVIEW AVE

(map) Reconstructed from Seattle
Commons Draft Plan, 1995.

(photos, below) South Lake Union, 2019.

REPUBLICAN ST

HARRISON ST

8TH AVE

9TH AVE

THOMAS ST

TERRY AVE

BOREN AVE

DENNY WAY

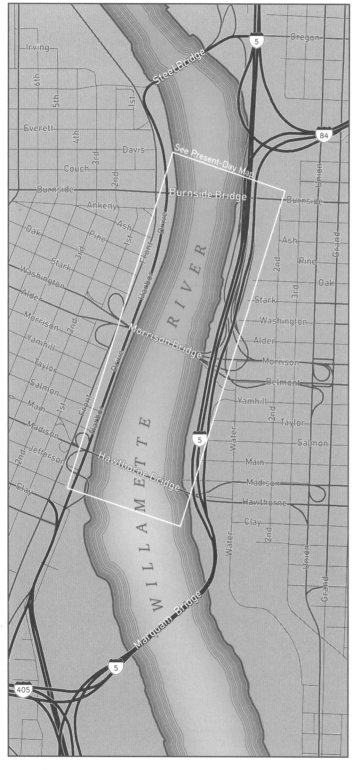

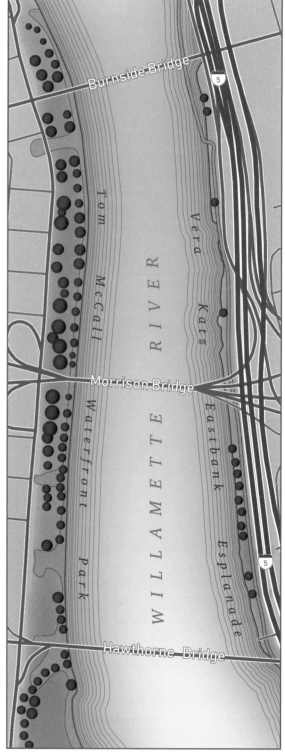

THE HIGHWAY AND THE WATERFRONT PARK

Before Portland's Tom McCall Waterfront Park was named one of the American Planning Association's great public spaces, it was home to the Harbor Drive Freeway. Built in 1942, the Harbor Drive highway was a four-lane thoroughfare hugging the Willamette River from south of downtown to the Steel Bridge. At the time, only a narrow sidewalk lined with trees gave pedestrians refuge from the highway just steps away.

Advocacy to convert the area to public space emerged during the 1960s when a freeway revolt was going on in cities across the country. In 1969, the State Highway Department proposed to add more lanes to Harbor Drive. An anti-freeway group, Riverfront for People, responded by hosting a picnic on the green expanse between Harbor Drive and Front Avenue. Two hundred and fifty adults and one hundred children attended the picnic and held signs that read "Parks for People" and "Save Our Riverfront."

Oregon's politicians, including Governor Thomas McCall, sided with the protesters. They rejected the proposal to enlarge Harbor Drive and instead decided to transform the space into Portland's thirty-six-acre pedestrian-focused Waterfront Park. The completion of I-5 also made Harbor Drive obsolete. The 1974 removal of Harbor Drive is viewed as a milestone in the US anti-freeway movement; it was one of the first cases of getting a pre-existing freeway removed (previous victories had come from blocking construction of new freeways). Remnants of Harbor Drive are still with us, including the southwest segment and the bicycle- and pedestrian-only ramp exiting the Hawthorne Bridge.

The Waterfront Park was named for McCall in 1984. The development cost approximately $20 million and rolled out in five phases across seventeen years. As part of the plan, designers competed to reimagine a ten-acre parcel in the South Waterfront. This resulted in RiverPlace, a private development including a public marina, rental housing, condominiums, retail and office space, an athletic club, and restaurants. Private development is also found to the north of Waterfront Park, where McCormick Pier Condos nestle up to the waterfront beyond the Steel Bridge.

On the east side of the Willamette, a different story prevails. Waterfront Park visitors can see the traffic on I-5 as it sweeps across the east bank. Riverfront for People has been advocating against freeway expansions on the east bank since the late 1980s—a charge recently picked up by the No More Freeway Expansions Coalition in their battle to challenge plans for widening the I-5 in the Rose Quarter. Will history repeat itself? Keep your eye out for picnickers with protest signs.

(right) Harbor Drive looking north toward Steel Bridge, 1950.

GOODBYE, EMBARCADERO FREEWAY

Since it was proposed to the city of San Francisco, the Embarcadero Freeway has been passionately opposed. Spanning from Second and King Street under the San Francisco–Oakland Bay Bridge to Pier 45, the double-decker Embarcadero Freeway famously blocked views of the waterfront for three decades. As one journalist remarked, "no one misses it."

When construction began in the 1950s, many were opposed to it since it would block the view of the Ferry Building and the San Francisco Bay. During a lull in construction from 1959 to 1964, thousands of anti-freeway protesters rallied to speak out against several freeway expansion plans in San Francisco. Their efforts effectively ended construction of the freeway. This halt in construction left a stunted and unfinished "freeway to nowhere" plaguing the view corridor. Despite the passionate dissent, businesses in Chinatown and other adjacent business districts relied on the freeway to usher customers to their doorsteps. Such claims held weight among decision makers in the city, which allowed for the defeat of a proposal to tear down the freeway in 1987.

The Loma Prieta earthquake of 1989 opened a window of opportunity for anti-Embarcadero Freeway proponents. The city found that the cost of reconstructing the damaged Embarcadero Freeway was greater than the cost of dismantling the freeway and redeveloping the space. In 1991, the city said its final farewell to the Embarcadero Freeway. Initially, the removal brought congestion, but increases in alternative transportation options eased traffic. By 2002, the redeveloped multiuse boulevard spanned six lanes of tree-lined streets, green bike paths, transit lines, and pedestrian paths. These transportation options have proven to be a boon for development along the boulevard, as businesses have taken advantage of the increased foot traffic and land values have increased (by some estimates as much as 300 percent).

The spaces where the freeway was removed have largely been converted into green open spaces. Sue Bierman Park (formerly known as Ferry Park), Rincon Park, Children Park, and the Vaillancourt Fountain now fill much of the Embarcadero space where the freeway once stood.

Down Folsom Street, the SoMa (South of Market) green space along Spear Street is a privately owned green space among many others tucked away between towers. The waterfront itself now offers a promenade and a public plaza. Where off-ramps once adorned the view corridor, the headquarters of the Gap now protrudes.

Construction of the Embarcadero Freeway, San Francisco, 1957.

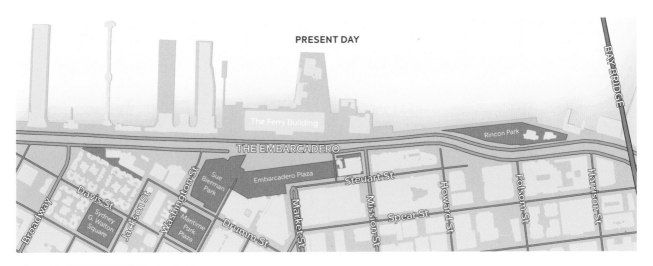

PRESENT DAY

THE EMBARCADERO

The Ferry Building

Rincon Park

Embarcadero Plaza

Steuart St

Broadway

Davis St

Sydney G. Walton Square

Jackson St

Washington St

Sue Bierman Park

Maritime Park Plaza

Drumm St

Market St

Mission St

Spear St

Howard St

Folsom St

Harrison St

BAY BRIDGE

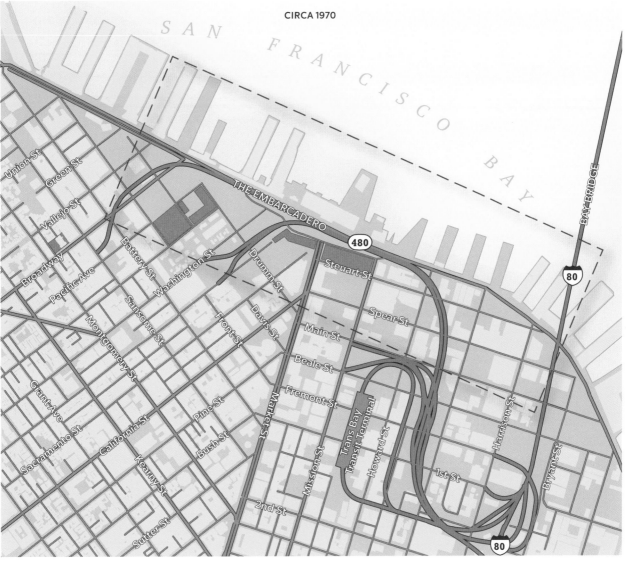

CIRCA 1970

S A N   F R A N C I S C O   B A Y

THE EMBARCADERO

480

80

BAY BRIDGE

Union St

Green St

Vallejo St

Broadway

Pacific Ave

Battery St

Washington St

Drumm St

Steuart St

Spear St

Sansome St

Front St

Davis St

Main St

Beale St

Montgomery St

Grant Ave

Sacramento St

California St

Kearny St

Pine St

Bush St

Fremont St

Market St

Mission St

Trans Bay Transit Terminal

Howard St

Harrison St

1st St

Bryant St

Sutter St

2nd St

80

SKYLINES

An urban skyline is a city's horizon—the boundary dividing the terrestrial environment from the sky. Each skyline is distinct, the loftiness of the tallest skyscrapers, range of building heights, and distinctive works of architecture. However, skylines change. Things appear or disappear from a city's silhouette and the horizon shifts.

The vantage point of our Seattle skyline is West Seattle. For Portland, this is the view from the Eastbank Esplanade. For San Francisco, our vantage point is Treasure Island.

San Francisco's skyline is in part distinctive because of the Transamerica Pyramid, the tallest building in the city until 2018, when the massive

sixty-one-floor Salesforce Tower opened, becoming the only skyscraper in any of the three cities more than one thousand feet tall.

Portland's skyline is the least distinctive in that the skyscrapers are of relatively similar size. This renders the Portland skyline less varied than the skylines of Seattle and San Francisco. The second-largest building in the city, the US Bancorp Tower, is called Big Pink locally—remarkable more for its color than any other reason.

Seattle's skyline is distinguished by the Space Needle, a remnant from the Century 21 Exposition of 1962. The Space Needle was built to associate the city with progress. The structure's location is apart from some of the city's other tall structures and gives contrast to the skyline.

GRAVEYARD SHIFT

Room for the dead is running out. Urban cemeteries are becoming increasingly rare, and those in the Upper Left are no exception. San Francisco, Portland, and Seattle have seen the vanishing of cemeteries and the exhumation of the dead themselves.

Across the world, burial is a common way to both dispose of bodies and provide a space for commemoration. American cemeteries grew out of the Christian practice of delineating and personalizing grave sites. Mid-nineteenth-century European Americans generally buried their dead in communal graves, churchyard cemeteries, small individual sites, or family plots. By the 1890s larger cemeteries of about 250 graves were common throughout much of the United States. In the twentieth century, cemeteries with hundreds or thousands of graves became the norm.

SAN FRANCISCO

San Francisco has only three cemeteries, the Mission Dolores Cemetery, the San Francisco National Cemetery in the Presidio (which technically isn't even part of the city), and a remarkably well-preserved pet cemetery (also in the Presidio). As the map indicates, there used to be more cemeteries in town, including the "Big Four" around Lone Mountain—Laurel Hill, Odd Fellows, Masonic, and Calvary.

(top) Workmen digging up graves in Odd Fellows Cemetery, San Francisco. September 28, 1931.

(right) San Francisco National Cemetery, 2019.

By the late eighteen hundreds, San Francisco cemeteries had reached capacity, were largely overgrown and falling apart, often served as a site of criminal activity, and were perceived to be a health hazard. Developers also wanted this prime real estate in the rapidly growing city. In the 1890s, the Jewish cemeteries closed and were developed into Mission Dolores Park. A new Jewish cemetery was established in the farmland of what is now the town of Colma.

San Francisco banned burials within city limits after August 1, 1901. Just over a decade later, in 1913, the Board of Supervisors went a step further and ordered the closing of all cemeteries and mandated removal of all bodies. By the 1940s, the remains of 150,000 people were disinterred and moved from the Big Four out of town.

The bodies went to Colma, a town incorporated in 1924 by an association of cemetery owners. It now has nearly twenty cemeteries and a living population of approximately 1,500. It is said that the dead outnumber the living by 1,000 to 1 in Colma.

The abandoned Laurel Hill Cemetery was subsequently developed into residential blocks. The Masonic Cemetery became the site of the University of San Francisco. A public golf course and the California Palace of the Legion of Honor now sit on the site of City Cemetery. A 1993 renovation of the latter yielded more than seven hundred bodies remaining from the old cemetery.

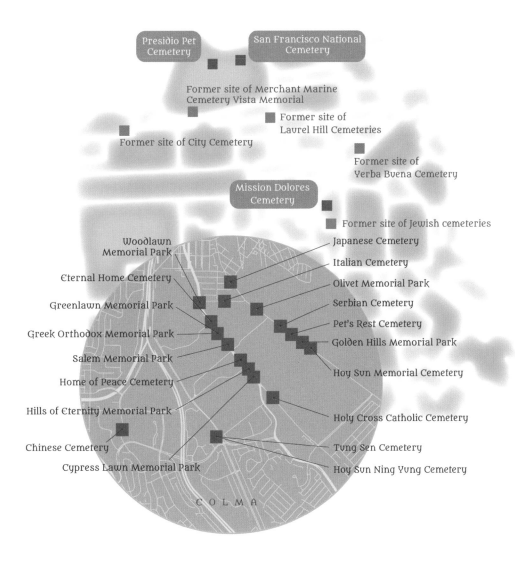

SEATTLE

White settlers' first cemetery in Seattle was open from 1853 to approximately 1860, and located on land that may have been a Native American burial ground. The Denny Hotel, downtown at Second Avenue and Stewart Street, later occupied that site, so the cemetery became anachronistically referred to as the Denny Hotel Cemetery. The remains of settlers and Native Americans are likely under the Moore Theatre and the Josephinum Apartments that stand there today.

The first official cemetery in the city, the Seattle Cemetery, is the present-day site of Denny Park, just north of downtown. Around 1861, bodies from informal sites were moved to the Seattle Cemetery to make room for development. By 1884, the cemetery was closed and more than two hundred bodies were moved to other cemeteries, mostly the Seattle Masonic Cemetery (currently named the Lake View Cemetery), to make room for the park. Circa 1930, as the site was lowered some sixty feet for the Denny Regrade, unearthed bodies were removed at night so as not to cause alarm.

The Seattle Masonic Cemetery was established in 1872 and eighteen years later renamed the Lake View Cemetery. It is the resting place of many notable historical figures, including Arthur Denny, "Doc" Maynard, Henry Yesler, and Princess Angeline—the eldest daughter of Chief Seattle. Martial arts icon Bruce Lee and his son, Brandon Lee, are also buried here side by side.

Comet Lodge Cemetery, located southeast of Beacon Hill, served as a sacred burial ground for the Duwamish before the Independent Order of Odd Fellows took title to the land and established a cemetery at the site in 1895. Largely abandoned by the 1930s, the cemetery became overgrown and forgotten. In 1987, the city of Seattle bulldozed the area and many grave sites for residential development. Hundreds of bodies are still believed to rest under the development. Gravestones that survived the bulldozers now rest randomly scattered rather than over actual grave sites.

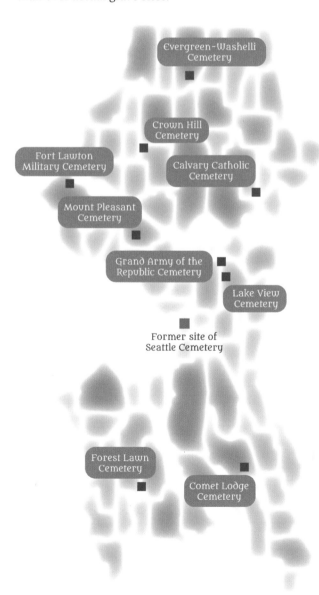

Plan for Denny Park on the site of the Seattle Cemetery, 1884.

PORTLAND

Portland has kept its formally established urban cemeteries, allowing them to both nestle into residential areas and be shoehorned by busy roads.

Mount Crawford Cemetery was established in 1855, but the first burial on the site took place in 1846. A remote and rural setting on private land far from Portland, the burial grounds were renamed Lone Fir Cemetery in 1866 for a solitary tree on the site. That fir tree is still there, but it is no longer alone; it is now part of the city's second-largest arboretum.

Circa 1855, Portland closed its first graveyards located in and around downtown at SW Ankeny and Front, SW Washington and SW Stark at Tenth, and SW Burnside and Eleventh. Bodies were moved to the relatively drier and well-drained confines of Lone Fir Cemetery, which is now on the National Register of Historic Places. Many graves are unmarked and more than twenty-five thousand people are buried there. The southwest corner of the cemetery was a section reserved for Chinese immigrants and asylum patients. The area was nearly developed in 2004, but bodies were discovered and the land reincorporated into the cemetery.

Portland's vast River View Cemetery in the southwest of the city was established in 1882 and is so large that more than half of it has not been developed as cemetery grounds. The city of Portland plans to maintain some of the area as natural space.

ON THE WATERFRONT

The 1954 drama *On the Waterfront* is a fictionalized account of shady mob-related criminal activities on the Hoboken Docks in New Jersey. Starring Marlon Brando, Eva Marie Saint, and Karl Malden (before he hit *The Streets of San Francisco*), the film promoted the idea that waterfronts were sketchy, crime-ridden areas where kickbacks, brawls, and murder were the order of the day. From this movie we get the notorious Brando-delivered line, "I coulda been a contender."

The contemporary waterfronts of San Francisco, Portland, and Seattle don't quite live up to this dramatic legacy. Remnants of the industrial grit still exist, however; as industry is increasingly replaced by green spaces, spendy residences, and marinas, the image of the waterfront has also shifted.

Downtown areas developed around a spot on the water that made it easy to conduct trade. Today each city has stretches of industrial waterfront, people living on the water in houseboats, and open recreational space.

Twenty-six miles of Portland's 117-mile city boundary is waterfront along the Willamette and Columbia Rivers. Eighty percent of Seattle's eighty-mile boundary is waterfront on Puget Sound and Lake Washington.

An impressive 88 percent of San Francisco's city boundary (including Treasure Island) is water. Ruins of the Sutro Baths on the ocean side rest in curious contrast to massive new developments on the bay side.

PORTLAND

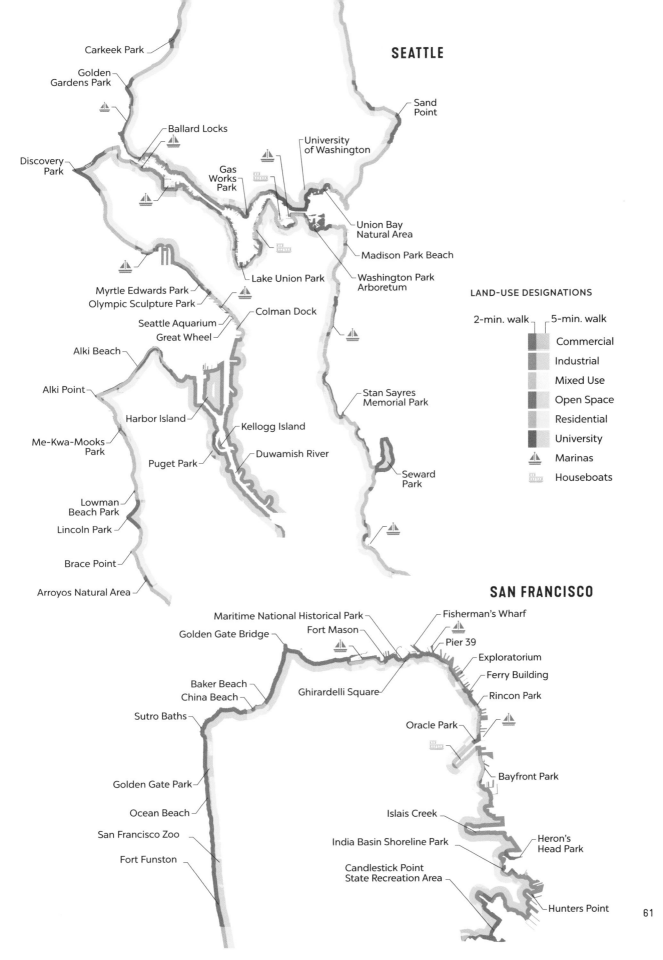

SEATTLE

Carkeek Park

Golden Gardens Park

Sand Point

Ballard Locks

University of Washington

Discovery Park

Gas Works Park

Union Bay Natural Area

Madison Park Beach

Lake Union Park

Washington Park Arboretum

Myrtle Edwards Park

Olympic Sculpture Park

Seattle Aquarium

Colman Dock

Great Wheel

Alki Beach

Alki Point

Stan Sayres Memorial Park

Harbor Island

Kellogg Island

Me-Kwa-Mooks Park

Puget Park

Duwamish River

Seward Park

Lowman Beach Park

Lincoln Park

Brace Point

Arroyos Natural Area

LAND-USE DESIGNATIONS

2-min. walk   5-min. walk

Commercial

Industrial

Mixed Use

Open Space

Residential

University

Marinas

Houseboats

SAN FRANCISCO

Maritime National Historical Park

Fisherman's Wharf

Golden Gate Bridge

Fort Mason

Pier 39

Exploratorium

Baker Beach

Ferry Building

China Beach

Ghirardelli Square

Rincon Park

Sutro Baths

Oracle Park

Golden Gate Park

Bayfront Park

Ocean Beach

Islais Creek

San Francisco Zoo

India Basin Shoreline Park

Heron's Head Park

Fort Funston

Candlestick Point State Recreation Area

Hunters Point

URBAN MOSAICS

Many of the maps in this book were made with quantitative data. But there are ways of understanding cities beyond numbers and statistics, ways more to do with paying close attention to your senses and emotions. Psychogeography is an approach to understanding how people experience places. It encourages us to pay explicit attention to what we hear, smell, touch, taste, and see in order to gain more nuanced understandings of cities. The maps on these pages were exercises in psychogeography.

It is difficult to represent geographies of experience. Unlike making maps of city streets or average rainfall, there are no standard conventions for mapping the way people feel about places. One thing we can do is pay attention to the textures of cities. Textures are hard and soft, sharp and round, rough and smooth.

To document our feelings about cities we can also focus on color. Colors have meanings; some meanings are culturally situated and others very personal. Throughout this book we occasionally use a red/green/blue color scheme for San Francisco, Portland, and Seattle. What was our thinking here? It was largely based on a general feeling, but we did recognize certain associations. . . .

Red is (sort of) the color of the Golden Gate Bridge. It is the representative color of the heart that one leaves in San Francisco. The 49ers wear red.

Green is the color of trees and plants, which works for mossy Portland, home to Forest Park. This is a city where grass grows through concrete and blackberry thickets run rampant. The Timbers wear green.

Blue is the color of water (not really) and Seattle is not only surrounded by it but has water running through it as well. Blue evokes the rainy mist that often embraces the city. The Mariners and Seahawks wear blue.

Here we present the colors and textures of each city. To create these maps, we stopped focusing on the flat gray of the city streets and sidewalks, instead concentrating on other landscapes. As we walked around these cities, we photographed the red in San Francisco, the green in Portland, and the blue in Seattle.

SEATTLE

PORTLAND

SAN FRANCISCO

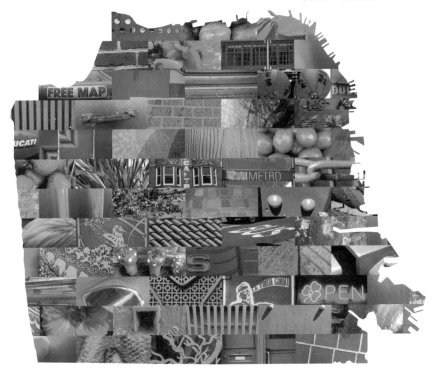

II. NATURE AND THE CITY

SAN FRANCISCO, PORTLAND, AND SEATTLE are known for their green spaces and nearby natural areas. San Francisco's Presidio and Golden Gate Park, Portland's Forest Park, and Seattle's Discovery Park all provide refuge from the hardscapes of the city. For all the mazes of concrete in these cities, there are still pockets of greenery and wildlife.

In San Francisco, Portland, and Seattle, the natural setting was formative. Water and waterways, such as natural harbors, were critical in the founding of each of these cities. From their founding to today, these cities' connections to nature are deep and wide—involving natural areas many miles away. As these cities grow, environmental goals come into tension with logistics, such as water sourcing and waste management.

These cities have a reputation for natural beauty. Ocean views (in San Francisco), bay views (in San Francisco and Seattle), and mountain views (in Portland and Seattle) provide residents and visitors with reminders of nature's presence in the urban landscape.

From trails to trees to cryptozoology, this chapter examines the interactions between the cities of the Upper Left and the natural world.

WHAT ARE YOU ON? LAND USE IN THE METRO AREA

Look around. What surrounds you? What is beneath your feet or outside your building? Is it asphalt, concrete, buildings, manicured lawns, an open field, a river, or perhaps a thick forest? Think about where you live. What covers the land? The type of land cover around you can impact your mood, your sense of place, and your general impression of an area. Land cover also reveals the primary use of an area, its history, its economic activity, and the way in which the landscape has (or hasn't) been modified.

On these pages, we explore the land cover makeup within 1, 5, 10, 25, and 50 miles of the center of downtown San Francisco, Portland, and Seattle. Looking at the dominant land cover as you move from the center of each city to 50 miles away, you can sense the character and feeling of each place captured in broad strokes.

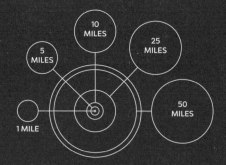

DEVELOPED
GREEN SPACE
FOREST
GRASSLANDS/SHRUBS
WETLANDS
OPEN WATER
AGRICULTURE

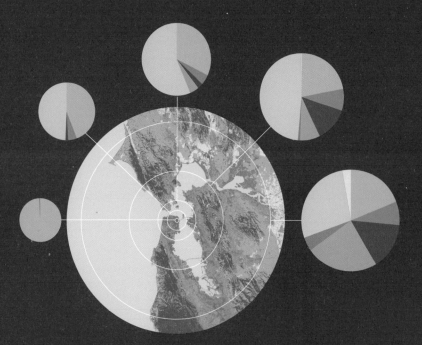

SAN FRANCISCO

1 mile: Concrete and asphalt dominate.

5 miles: Water is half of the overall coverage. Developed land is most of the rest, with green spaces and forests edging in.

10 miles: The ocean and bay make water the largest cover type. Developed space followed by green spaces, grasslands, and forests are apparent.

25 miles: Green land cover types increase. Water coverage is now less than half of total.

50 miles: Green spaces, forests, grasslands, and wetlands now make up more than half of the total land cover.

PORTLAND

1 mile: Mostly developed. Downtown, water (the Willamette River), and nearby green spaces make up most of this space.

5 miles: Still primarily developed. Water cover decreases. Green land cover types become prominent.

10 miles: Little changes between 5 and 10 miles from downtown.

25 miles: Increases in wetlands and grasslands and, most dramatically, in agriculture.

50 miles: Forest and agriculture dominate. Grasslands increase significantly, while developed land makes up the smallest percentage across scales for all three cities.

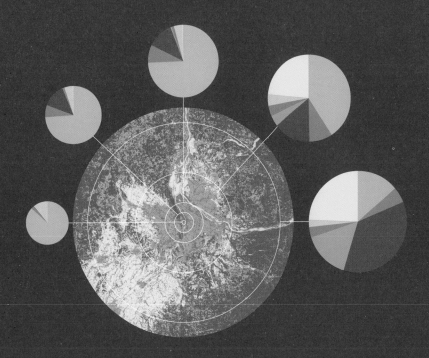

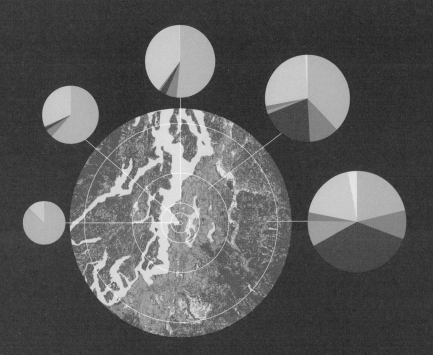

SEATTLE

1 mile: Urban development and water dominate the landscape in downtown Seattle.

5 miles: Developed land remains dominant, but water becomes more significant. Green spaces and forests emerge.

10 miles: Water continues to increase—more like San Francisco than Portland. Green spaces and forests are joined by slivers of grasslands and agriculture.

25 miles: Forest cover expands dramatically, and green spaces double. Water and development decrease.

50 miles: Forests are now the largest category. Grasslands and agriculture are more noticeable, while developed land recedes.

CREATURES OF HABITAT

The map on the adjacent page features representative animals from across the Upper Left. The map is not comprehensive—it features some of the animals that lend the Upper Left its special associations. Often these associations run deep (orcas). Sometimes the associations are highly localized (California newt) and sometimes they are more general (elk). We left out common animals that are found widely elsewhere (coyotes, crows, and deer).

On the map, we placed animals in locations where they are highly visible, locations with cultural ties to the animal, and locations where the animals once lived (grizzly bears).

CHINOOK SALMON
Prized fish of humans, bears, and sea lions alike.

NORTHERN SPOTTED OWL
Single-handedly transformed the logging industry.

ORCA
These apex predators are more closely related to dolphins than whales. Killer!

BANANA SLUG
Beloved mascot, but not as tasty as a banana.

ELEPHANT SEAL
At five thousand pounds, bigger than your car but still not as heavy as an elephant.

CALIFORNIA NEWT
Poisonous and proud of it in a coat lined with safety orange.

GRIZZLY BEAR
On the flag (and only on the flag) in California and hidden in the mountains of Washington.

LION
If you missed the billboards, you weren't paying attention.

AMERICAN BEAVER
Namesake of suburbs and sports teams and featured on the Oregon flag.

SEA LION
From Fisherman's Wharf to the Bonneville Dam, they like to bark and strike a pose.

ELK
Tourists charge past clearly posted signs for dramatic photo ops.

HARBOR SEAL
The most widely distributed of the pinnipeds.

NORTHERN FLYING SQUIRREL
They can fly! Not really.

PUFFINS
I think that's enough fish, buddy.

HUMPBACK WHALE
Every day is hump day.

GRAY WOLF
Once plentiful then driven out, they are beginning to return to the area.

A CAPTIVE AUDIENCE: ZOOS

A short drive or bus ride is all it takes for many to find themselves nose to nose with a lion (separated by alarmingly thin glass). It wasn't always so easy to see an animal from across the world so close to home. Zoos became popular across the country at the turn of the twentieth century. Zoos on the West Coast usually developed haphazardly. Private donations of one or two animals were often the impetus for cities to build an official collection.

Zoos were developed in the same spirit, and sometimes in the same location, as parks. Early developers thought of zoos as public institutions for middle-class urbanites. Zoo planners wanted to provide a "wholesome" break from urban living by connecting people with nature. Zoos are middle landscapes—offering a manicured and conveniently located reprieve from the sometimes grim realities of urban living.

But zoos are not without controversy. Early zoos often had ties to the circus and deeply misguided animal care practices. Given these missteps, it is easy to overlook the goals of early zoos to connect people with nature, further scientific study of animals, and develop animal husbandry techniques.

Early methods of deciding which animals to include in a zoo were motivated by what was gifted, trendy, or convenient. Elephants were a particularly popular choice for fledgling zoos. Elephants became one of the most controversial zoo animals due to their intelligence, sociality, and large home ranges in the wild. Many zoos that acquired elephants phased them out, such as the San Francisco Zoo and Woodland Park Zoo, or expanded and improved living conditions, like the Oregon Zoo.

Far from their beginnings, zoos in the Upper Left now house extensive animal collections. The San Francisco Zoo has more than 250 species, the Woodland Park Zoo has about 320 species, and the Oregon Zoo has about 230 species. Each of these zoos participates in species survival plans—collaborative efforts among zoos and aquariums that use captive breeding to prevent the extinction of threatened or endangered animals. In each collection, there are animals that stand out—animals whose names are known by residents, who show up on advertisements, and whose passing is mourned citywide.

When you think of an early zoo, you might think of a sad animal in an austere cage with large metal bars. Many of the ethically questionable animal husbandry practices had to do with widespread inexperience, lack of knowledge, and misinformation about animal care. Animal husbandry practices have come a long way; today they are standardized by organizations, conferences, and whole disciplines dedicated to the subject.

There has been a general movement over the last century away from organizing zoos by taxonomy and toward arranging by ecology. Contemporary zoo animal arrangements are often aimed at landscape immersion. The goal is to create a place that reflects the complex relationships between animals, plants, and their environment. Woodland Park Zoo was a pioneer in landscape immersion exhibits in the 1970s, and it remains a leader in ecologically focused exhibits.

Priorities have changed for zoos. Early zoos aimed to civilize and enrich urban populations. Conservation, education, and research are now prioritized along with recreation. Current zoos function as places to garner support from the public for animals and ecosystems worldwide. Zoos continue to bring animals and people together and fuel conversations about animal rights, conservation, and biology.

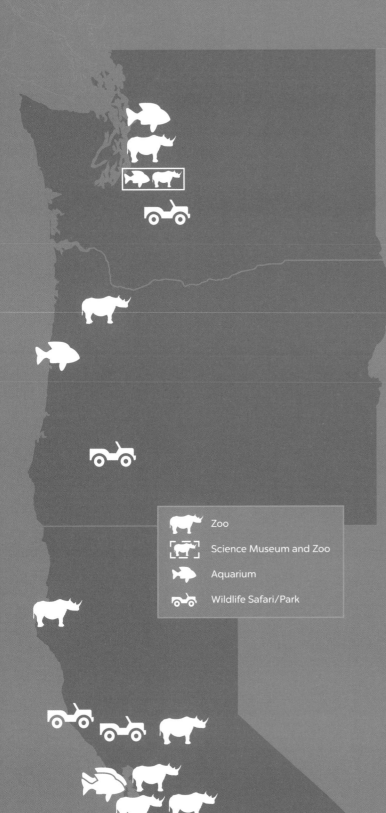

Legend

- Zoo
- Science Museum and Zoo
- Aquarium
- Wildlife Safari/Park

(top) Deer, Woodland Park Zoo, Seattle, 1914.

(center) Bear Pit, Portland Zoo, 1935.

(bottom) Children and elephants, San Francisco, Fleishhacker Zoo, 1949.

DISTINCTIVE ANIMALS

SEATTLE

DEER
Guy Phinney, a real estate developer, created a deer park on his estate. When his estate became part of Woodland Park in 1889, so did the deer. A herd of buffalo soon followed.

GORILLAS
Western lowland gorillas have been a fixture at Woodland Park since the 1950s. Bobo, a former family pet, was a main attraction of the zoo during his short life (1951–1968).

PORTLAND

GRIZZLY BEARS
A female grizzly bear gifted from a local pharmacist in 1888 prompted city government to establish the Portland Zoo (now the Oregon Zoo).

ASIAN ELEPHANTS
The Oregon Zoo features Asian elephants; most famously, Packy (1962–2017), the first elephant born in the Western Hemisphere since 1918 and one of the tallest and oldest elephants in the United States.

SAN FRANCISCO

GRIZZLY BEARS
Monarch, captured in 1889 and thought by some to be the last grizzly bear in California, inspired the development of the San Francisco Zoo. Monarch's likeness was later used for the California flag.

MAGELLANIC PENGUINS
The San Francisco Zoo has the world's largest captive colony of fidgety and fashionable Magellanic penguins. The birds hail from the Strait of Magellan at the southern tip of South America.

SEATTLE

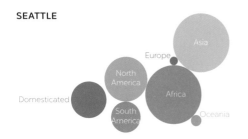

MAMMAL TOTAL

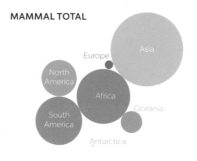

PORTLAND

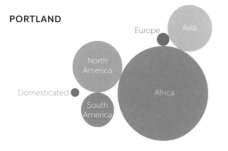

ANIMAL ORIGINS

Which continent's mammals are best represented at each zoo? These figures do not show where an individual animal came from, but show instead the home continent of each species (in many cases, there is more than one). These figures show species diversity, so large collections of a single species do not impact the figures. The top figure provides a baseline—this is the overall proportion of mammals on each continent.

SAN FRANCISCO

ANIMAL ARRANGEMENTS

How are animals arranged in a zoo? These maps show each zoo's layout. You can see (roughly) how much space is dedicated to each section and how they connect. The Woodland Park Zoo is the only zoo in the group that has ecological sections. The Oregon Zoo has a large section devoted entirely to Asian elephants. The San Francisco Zoo has the most sections still categorized by taxonomy. Notice that both the Oregon Zoo and San Francisco Zoo have concentrated activity areas (primarily aimed toward children, with play structures and seating).

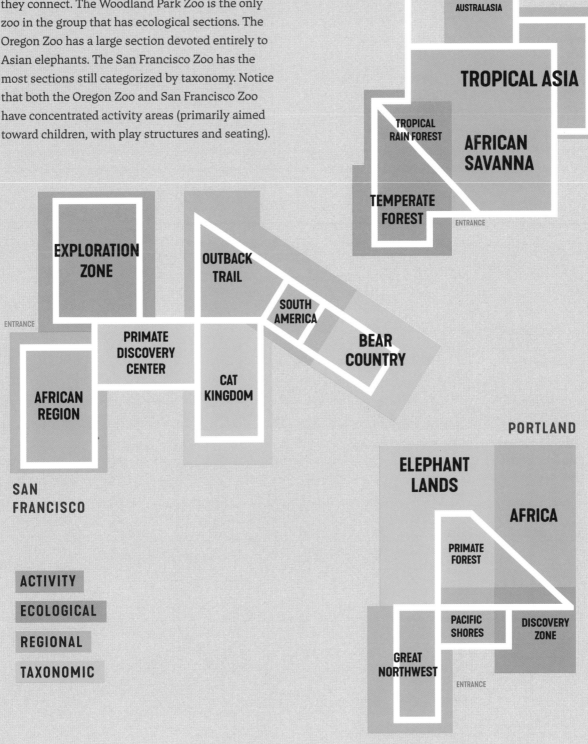

SEATTLE

NORTHERN TRAIL

AUSTRALASIA

TROPICAL ASIA

TROPICAL RAIN FOREST

AFRICAN SAVANNA

TEMPERATE FOREST

ENTRANCE

EXPLORATION ZONE

OUTBACK TRAIL

SOUTH AMERICA

BEAR COUNTRY

ENTRANCE

PRIMATE DISCOVERY CENTER

CAT KINGDOM

AFRICAN REGION

SAN FRANCISCO

ACTIVITY
ECOLOGICAL
REGIONAL
TAXONOMIC

PORTLAND

ELEPHANT LANDS

AFRICA

PRIMATE FOREST

PACIFIC SHORES

DISCOVERY ZONE

GREAT NORTHWEST

ENTRANCE

TRAILS AS TRANSIT MAPS

Public transit often defines a city. Think about the classic London, Paris or New York City transit maps. Broad lines of varying colors crisscross the page in a way that's easily understandable, yet completely unlike the geography of the actual city. Their purpose is to relay information about a transit network in the simplest way.

Trails are not unlike public transit lines. They enable pedestrians and bicyclists to get places they might not otherwise be able to access. Trails are often hidden in maps that are complicated to read or buried beneath the road network. But what if trails were visualized in the same way that transit was—with an easy-to-read and understandable map? The maps of regional trails on these pages attempt to do just that.

PORTLAND

The metro region of Portland has an extensive network of trails thanks to Metro, its regional planning agency. Metro spearheads the planning and development of trails through its tri-county area. Clark County, Washington (just north of Portland), also contributes to this ever-growing trail network. To date, the Portland metro region is home to more than three hundred miles of connected regional trails, most of which are hard surface that allow for easy access for those who bike, walk, or roll. Utilizing Portland's trail system, a person could walk from Vancouver, Washington, through downtown Portland, and out to Hillsboro almost entirely off street.

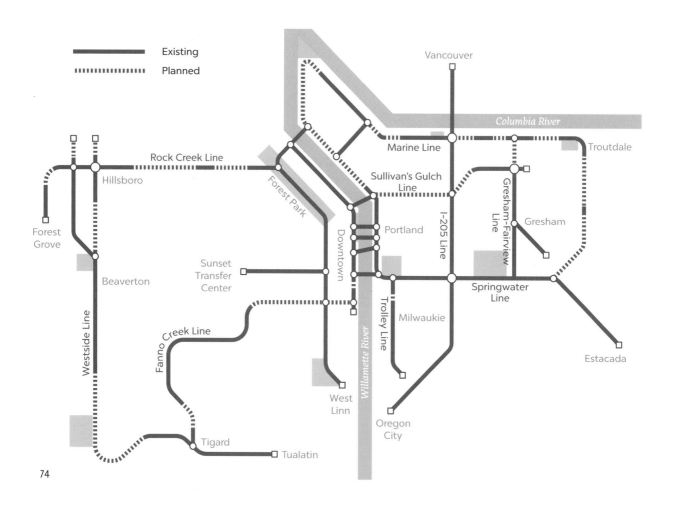

SEATTLE

King County is the primary trail provider for the Seattle metro region, though Pierce County and Snohomish County also have their own regional trails. King County has more than three hundred miles of paved and unpaved trails throughout the county. The county itself manages the planning and development of about half of those trails, with individual cities in the county managing the other half. While Seattle's regional trail system isn't managed quite as holistically as Portland's, King County has goals to eventually connect their trail network with the Puget Sound region by 2050, which will effectively double the amount of trail miles for the region.

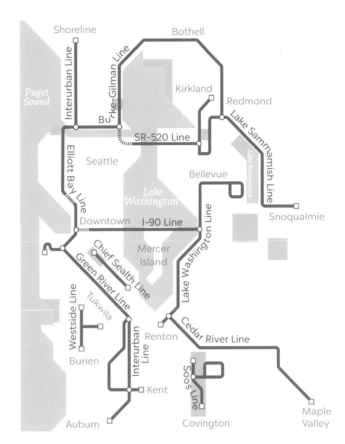

SAN FRANCISCO

Bay Area trail development is less centralized. That said, the Association of Bay Area Governments does manage the development and planning of the Bay Trail, a five-hundred-mile trail that wraps all the way around the bay, from San Jose to the south all the way through San Francisco and up to Vallejo in the north. Like the Bay Trail, the Bay Ridge Trail completes a large loop around the entire San Francisco and San Pablo Bays. Unlike the Bay Trail, however, the Ridge Trail is mostly soft surface, meaning it's largely for hiking and recreation, rather than providing transportation connections across the region.

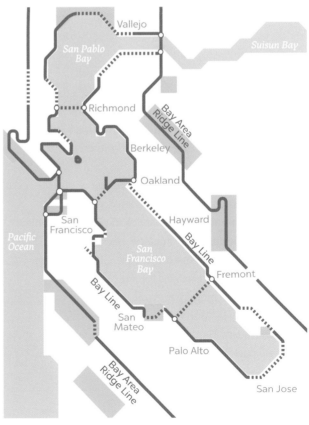

DRINKING WATER

San Francisco, Portland, and Seattle are all located on bodies of water. But where does the drinking water for these cities come from . . . ?

SAN FRANCISCO

San Francisco gets its water from a complex water supply system connecting the Sierra Nevada to the city through a series of reservoirs, tunnels, pipelines, and treatment systems. The Hetch Hetchy Watershed, located more than 180 miles away in Yosemite National Park, provides approximately 85 percent of San Francisco's water, while the Alameda and Peninsula Watersheds provide most of the rest. The Calaveras and San Antonio Reservoirs in the Alameda and Peninsula Watersheds capture and store rain and runoff. Since 1950, the majority of San Francisco's water has been fluoridated.

Lake Pilarcitos, a small reservoir still in use today, was the first year-round source of water for the rapidly growing city. Pilarcitos was owned by the Spring Valley Water Company, a private company that had monopoly control over San Francisco's water supply until the early twentieth century. In 1930, San Francisco purchased the Spring Valley Water Company and gained control of its water supply.

From 1901 to 1913, environmentalists were locked in a bitter battle with federal, state, and local government over the fate of the Hetch Hetchy Valley. The city of San Francisco proposed that a dam be built on the Tuolumne River (located in the valley) to create a reservoir for a reliable source of water and ultimately prevailed. Today Hetch Hetchy Valley is buried under water behind the O'Shaughnessy Dam. The damming of water on national park land is contested by environmentalists to this day.

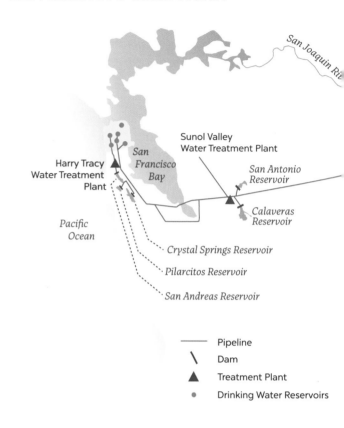

Pipeline
Dam
Treatment Plant
Drinking Water Reservoirs

SEATTLE

Seattle gets almost all of its water from two watersheds fed by the Cascades. The Cedar River Watershed is the largest source of water for the city, providing about 65 percent of their water, and is located about thirty miles from Seattle. Water flows from Chester Morse Lake into Cedar River, which leads it to Lake Youngs, where it enters Seattle's pipeline system.

The South Fork Tolt River Watershed, located in the foothills of the Cascades in east King County, provides Seattle with the rest. This watershed

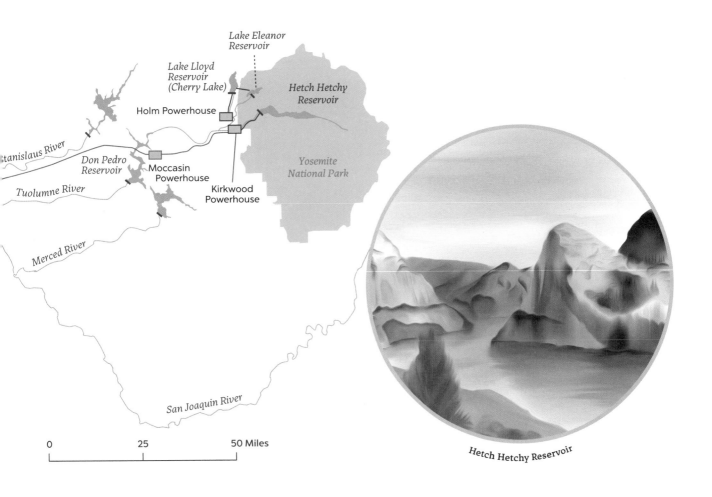

Hetch Hetchy Reservoir

drains into the Tolt Reservoir, and then is pumped into municipal pipelines. Both of the watersheds are owned by the city of Seattle.

Seattle has been covering its reservoirs since 2004—underground structures and ninety acres of public space have replaced six open reservoirs. Bitter Lake and Lake Forest Park Reservoirs have been fitted with floating covers and the remaining two open reservoirs have been decommissioned. Since 1970, Seattle has been fluoridating its water.

In the 1880s, the city was still pumping its water from nearby springs and lakes. Private companies monopolized the water market, drawing water from Lake Washington and Lake Union and selling it to city residents at a premium. The aftermath of the 1889 fire confirmed the inadequacy of existing water systems. The city established a public-owned municipal waterworks whose primary source would be the Cedar River Watershed.

Despite the near-constant rain in the city, Seattle actually has one of the highest water bills in the country. The high rates are due in part to significant investments in water infrastructure. The city recently spent millions to relocate their reservoirs underground to comply with EPA clean standards.

PORTLAND

Most of Portland's water comes from the Bull Run Watershed, located about 30 miles east of Portland in the Sandy River Basin. The protected watershed collects water from rain and snowmelt and has been the primary source of drinking water for the Portland area since 1895, when it began to flow to the city through 24 miles of pipelines constructed through old-growth forest. The Columbia South Shore Well Field, established in 1984, is used as a secondary source of water.

Portland is the largest city in the United States not to fluoridate its water. Though ballot measures to add fluoride to the city's drinking water have come up for a vote four times between 1956 and 2013, residents have consistently decided not to fluoridate. The issue remains polarizing among city residents.

Until 2015, Portland stored much of its drinking water in uncovered, open-air reservoirs, three at Mount Tabor Park and two in Washington Park. The reservoirs are perhaps a little too open-air. In 2008, two skinny-dippers were caught frolicking in a reservoir. In 2011, the city of Portland had to dump thirty-eight million gallons of drinking water after a nineteen-year-old was spotted urinating into a Mount Tabor reservoir.

In 2006 the EPA passed legislation that put a stop to uncovered reservoirs in the United States to prevent the spread of the microorganism cryptosporidium. Reluctant to cover their reservoirs, in 2015 Portland began the process of replacing its existing open drinking water facilities with enclosed storage and disconnecting the uncovered reservoirs from the drinking water system. While the old uncovered reservoirs no longer supply drinking water, the city of Portland continues to maintain and preserve them, as important community and historical assets.

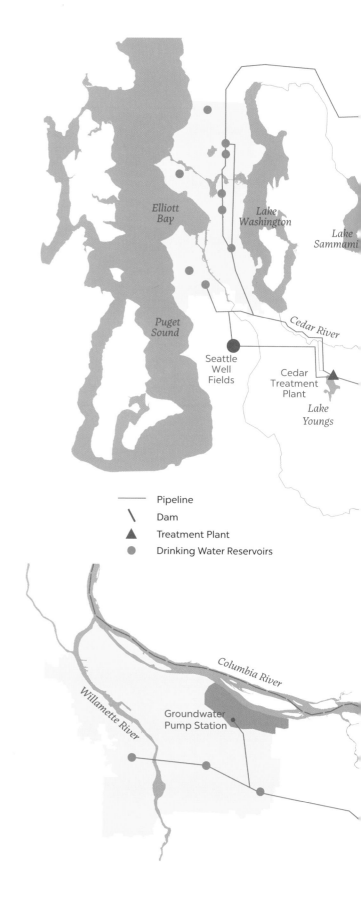

- —— Pipeline
- \ Dam
- ▲ Treatment Plant
- ● Drinking Water Reservoirs

SEATTLE'S WATER SYSTEM

Tolt
Treatment
Plant

South Fork Tolt Reservoir

Tolt River

Tolt River Watershed

Snoqualmie River

0   10   20 Miles

Landsburg
Diversion
Dam

Masonry
Pool

Chester Morse Lake

Cedar River Watershed

Green River

PORTLAND'S WATER SYSTEM

WASHINGTON

OREGON

Reservoir 2

Reservoir 1

Bull Run Watershed

Bull Run Lake

ly River

South Fork Tolt Reservoir

Chester Morse Lake

Bull Run Lake

CONCRETE JUNGLE

Trees soften harsh urban environments. We fill our parks with trees. Yet how much of our day-to-day life do we spend in a park? Parks and their trees are few and far between, whereas streets, and by extension street trees, run all over the city. When looking at these maps, it is important to remember that they show only street trees, not all the trees in the city, which would include trees in parks and on private property.

The trees that cities choose to plant reflect different landscape aesthetics and values. Early on, street trees were chosen primarily for their ornamental value. Today the criteria rests more on

fitting trees into available space, which is largely driven by avoiding power lines. Disease/pest resistance, canopy diversity, and known success in an urban environment are also factors.

Most street trees are non-native. Europeans brought many of their favorite trees with them to re-create the landscapes of the homes they left behind. Many of the oldest and wealthiest areas in each city have a high number of street trees and high diversity of tree species.

In some cases, native trees don't make great street trees. The Douglas fir, native to Portland and Seattle, grows quite tall. Some are found as

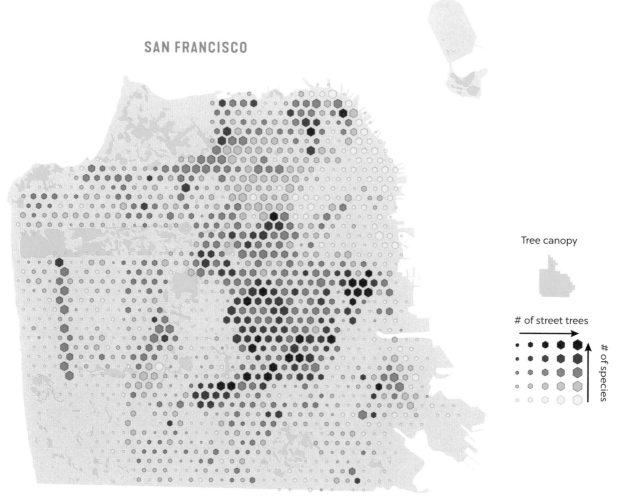

SAN FRANCISCO

Tree canopy

# of street trees

# of species

street trees, largely in the outer southwest and eastern areas of Portland and the northern area of Seattle. The Monterey cypress, native to the central California Coast, gets large and sprawling, although Sunset, Geary, and Junipero Serra Boulevards have a bunch.

The non-native red maple and Norway maple are among the most numerous street trees in both Seattle and Portland. San Francisco has more than eleven thousand gum trees, including eucalyptus, a non-native maligned for its propensity to spread fire over wide areas.

Street trees are also related to sidewalks. In places where there are no sidewalks, there tend to be few or no street trees. Tree roots sometimes transform, if not mangle, sidewalks and provide their own input on the flow of people through the city.

Unlike Portland and Seattle, San Francisco has relatively few street trees and few species downtown. Seattle and San Francisco have more concentrated patterns of tree species in particular neighborhoods; Portland has more species all over the place. Unlike the other cities, San Francisco has several large arterials lined almost exclusively with one species of tree.

Portland has such a reputation for cutting down trees that one of its nicknames is Stumptown. When it comes to street trees, species in Portland are well distributed. Dense concentrations of

PORTLAND

Tree canopy

# of street trees

# of species

street trees are found in old streetcar suburbs. The east side of the city, which includes many neighborhoods annexed in the 1980s, has relatively few street trees (and, relatedly, few sidewalks).

Although North Seattle has many trees, including the giant native Douglas fir, it does not have many street trees (as the map shows). Once again, we find dense concentrations of street trees in some of the older and more established neighborhoods.

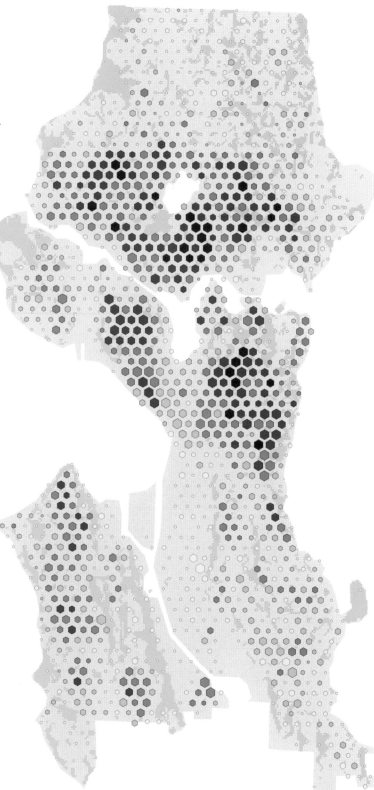

Tree canopy

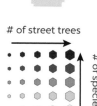
# of street trees

# of species

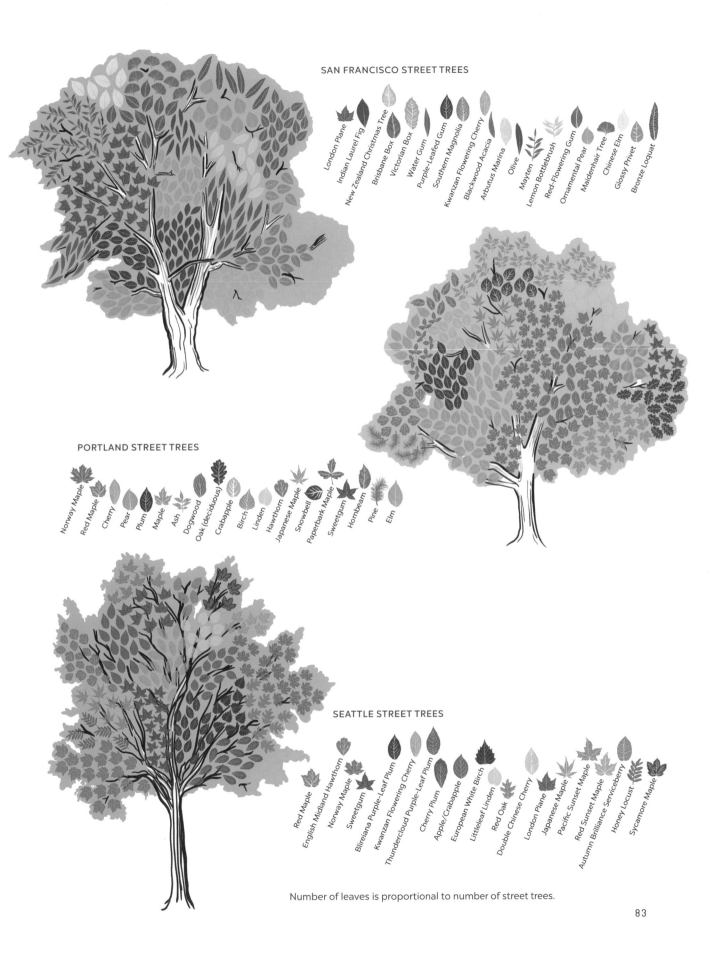

SAN FRANCISCO STREET TREES

London Plane · Indian Laurel Fig · New Zealand Christmas Tree · Brisbane Box · Victorian Box · Water Gum · Purple-Leafed Gum · Southern Magnolia · Kwanzan Flowering Cherry · Blackwood Acacia · Arbutus Marina · Olive · Mayten · Lemon Bottlebrush · Red-Flowering Gum · Ornamental Pear · Maidenhair Tree · Chinese Elm · Glossy Privet · Bronze Loquat

PORTLAND STREET TREES

Norway Maple · Red Maple · Cherry · Pear · Plum · Maple · Ash · Dogwood · Oak (deciduous) · Crabapple · Birch · Linden · Hawthorn · Japanese Maple · Snowbell · Paperbark Maple · Sweetgum · Hornbeam · Pine · Elm

SEATTLE STREET TREES

Red Maple · English Midland Hawthorn · Norway Maple · Sweetgum · Blireiana Purple-Leaf Plum · Kwanzan Flowering Cherry · Thundercloud Purple-Leaf Plum · Cherry Plum · Apple/Crabapple · European White Birch · Littleleaf Linden · Red Oak · Double Chinese Cherry · London Plane · Japanese Maple · Pacific Sunset Maple · Red Sunset Maple · Autumn Brilliance Serviceberry · Honey Locust · Sycamore Maple

Number of leaves is proportional to number of street trees.

NATURAL DISASTERS

Earthquakes, volcanoes, fires, floods—it's a wonder anyone lives in the Upper Left at all. Seattle is ranked number one in the country for the number of potential hazards. Emergency management policies, building codes, and community efforts are a response to potential threats. As hazards make their mark, the form of each city adapts. Doomsday preppers, this one's for you.

EARTHQUAKES

Upper Left natural hazards are connected to a belt of activity, the Ring of Fire, that encircles the Pacific Ocean with volcanic eruptions and tumultuous fault lines. Among the most active of these is the San Andreas Fault, which in 1906 triggered the deadliest earthquake in US history.

EARTHQUAKE IMPACTS SAN FRANCISCO, 1906 AND 1989

THE GOLDEN GATE

Chinese refugees segregated in north Presidio camp

THE EMBARCADERO

THE BAY BRIDGE

HUNTERS POINT

1906 EARTHQUAKE
Refugee camps
Area burned

1989 EARTHQUAKE
Damaged Embarcadero Freeway

CONTEMPORARY SF
Liquefaction zones
Tall buildings

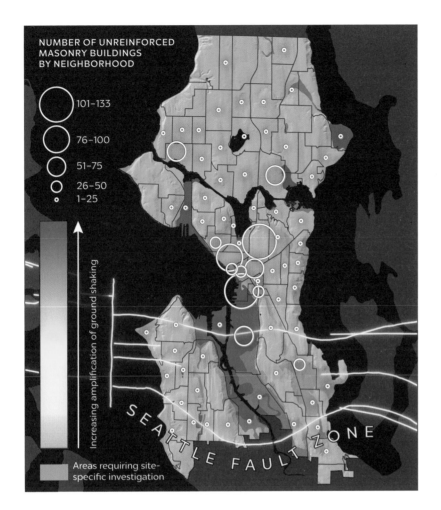

NUMBER OF UNREINFORCED
MASONRY BUILDINGS
BY NEIGHBORHOOD

101–133

76–100

51–75

26–50
1–25

Increasing amplification of ground shaking

Areas requiring site-
specific investigation

SEATTLE FAULT ZONE

San Francisco experienced a 7.7-magnitude event that caused buildings to collapse across the city. Broken natural gas mains and cutoff water supplies resulted in fires that burned for three days and destroyed five hundred city blocks, killing three thousand people and leaving more than a quarter of a million without homes. San Andreas shook again in the 1989 Loma Prieta earthquake, a 6.9-magnitude event televised live during the World Series at Candlestick Park.

While Seattle and Portland don't have movie-title-worthy fault lines, earthquakes are still regarded as high-risk hazards. Both cities are preparing intensely for a Cascadia Subduction Zone (CSZ) "megathrust" earthquake. The last Cascadia earthquake occurred in 1700, neatly following the CSZ's average of an earthquake every 350 years.

To plan for the worst, Seattle and Portland have started enforcing building code policies targeting their most vulnerable building types: unreinforced masonry. Among the list of beloved vulnerable structures, charismatic bridges are at high risk of collapse in a seismic event. In Portland, only the newly rebuilt Sellwood Bridge and the Tilikum Crossing are expected to withstand a megaquake.

Full-scale evaluations and citywide emergency planning are under continuous scrutiny. The 2016 Cascadia Rising exercise across the Pacific Northwest tested the response capacity of local and state agencies and volunteers. The exercise suggested that expectations of a seventy-two-hour recovery period, during which most people will need to be self-sufficient, needed to be extended to two weeks.

FIRES

San Francisco's 1906 fires weren't the first time the flames consumed the city, and some of these fires were not caused by natural forces. A December 1849 fire spread across a gold miners' tent encampment and was defeated only by destroying the remaining buildings in its path. This event catalyzed creation of the city's fire department, handy for the subsequent fires that would scorch the city, the greatest of which began on the evening of May 3, 1851. This fire started its way down Kearny, burning up the planked streets as it went and shifting with the hurricane-force winds. After ten hours of fire, eighteen city blocks of the main business district were destroyed. Merchants and property owners responded to the destruction by forming the first Committee of Vigilance. The committee soon found a street gang, the Sydney Ducks, to be responsible for the May destruction, setting fire to one part of the city in order to loot another.

Portland's most destructive fire was the Great Fire of 1873. Also suspected to be arson, the fire began in a furniture shop at First and Taylor and burned for twelve hours, allowing time for fire brigades from as far as Salem and Vancouver to arrive and help fight the flames. Twenty-two city blocks were destroyed. At the time, the city's fire bell was too weak to be heard over the sound of flames, so it was replaced after the incident.

Unlike these fires in San Francisco and Portland, Seattle's Great Fire of 1889 was not an intentional act, but rather an unfortunate

Panoramic view of earthquake and fire damage around Union Square (taken from the intersection of Stockton Street

RECENT DESTRUCTIVE FIRES IMPACTING AIR QUALITY

SEATTLE: 2018, 2020
Fire: British Columbia (2018)
Oregon (2020)

PORTLAND: 2017, 2018, 2020
Fire: Eagle Creek (2017)
British Columbia (2018)
Riverside (2020)
Beachie Creek (2020)

SAN FRANCISCO: 2018, 2020
Fire: Camp (2018)
LNU Lightning Complex (2020)

accident. A cabinet shop employee encountered a glue fire, which he tried to subdue using water. The mix of water and glue sent flames across the wooden room, soon consuming the cabinet shop. High winds met the flames and carried them over a twenty-five-block area. Unfortunately, the city's fire department was low on water due to low tide, which left few options available to fight the fire. Afterward, rebuilding processes included replacing wooden water pipes with more fire-retardant materials and using brick to reconstruct buildings.

Forest fires miles from a city can create dangerous air quality conditions in town. San Francisco, Portland, and Seattle have all had recent experience with this. In September 2020, smoke from massively destructive regional wildfires enshrouded each city. For up to two weeks, residents battened down windows and doors to ward off hazardous air quality and the omnipresent stink of smoke. For several days, Portland had the worst urban air quality on earth.

Volcanic Hazard Zones

VOLCANIC ERUPTIONS

If fires and earthquakes weren't enough, Portland and Seattle have giant volcanoes asleep in their backyards. Fortunately for Seattle, the two closest volcanoes, Mount Rainier and Glacier Peak, are too far afield for an eruptive blast to wound the city. The volcanoes map displays the historic paths of lahars (debris flows) that could emanate from volcanoes in Washington and Oregon. Neither Portland nor Seattle is in the direct path of lahars. Yet ashfall from any of Washington's five stratovolcanoes could cut off the city's transportation, power, and water supplies, thrusting Seattle into a volcanically induced isolation.

Mount St. Helens, sitting a bit closer to Portland than Seattle, is the most recently active volcano in the Cascades. Its May 18, 1980, eruption lasted nine hours, ejected ash into the stratosphere, choked the

Columbia River with mud, and killed fifty-seven people. Favorable winds spared the west side of the Cascades from much ash that day, but a smaller eruption later in October dusted Portland with ash. Mount Hood, Oregon's tallest mountain, looms large to the east of Portland. The dormant (hopefully) Mount Tabor is a volcano within the city limits.

San Francisco doesn't have to worry too much about volcanoes. Although the Clear Lake Volcanic Field is only ninety miles north of the city, its last eruption was about ten thousand years ago. Being so close, what are the chances that a city-dweller could see volcanic ash from this volcano in her lifetime? The chances are very good—assuming one is submerged in a therapeutic mud bath, made from said ash, at one of the many spas situated near the volcanic field.

FLOODS

The Upper Left is also subject to flooding. Early Seattle's efforts to control flooding through landfill, drainage, and the channeling of the Duwamish have proven largely successful. In December 2006, rainfall of about 2.17 inches in twenty-four hours caused landslides and flooding that damaged three hundred homes while pushing stormwater facilities beyond capacity. Despite dramatic occurrences like this, flooding is viewed as a relatively low-risk hazard in Seattle.

Portland, located at the confluence of two rivers, is much more acquainted with floods, with the most severe episodes occurring in the late eighteen hundreds. The 1876 flood reached a high-water mark of 25 feet, a record high until 1894, which crested at 33.5 feet. The 1894 flood reached inland as far as Northwest Tenth and Glisan Street, and Southwest Sixth and Washington Street.

Portland's most devastating flood did not have the highest water mark. The Vanport Flood of 1948 displaced an entire town. Developed to temporarily house wartime shipbuilders, at its peak Vanport was home to forty thousand workers. After the war, the town became one of the few locations African Americans could reside, given redlining and the city's efforts to keep neighborhoods segregated. The town was built on the Columbia River floodplain and surrounded by dikes up to a height of twenty-five feet. The onslaught of rain and melting snowpack pushed against and compromised the dikes, leaving Vanport completely inundated and displacing 18,500 residents, of which roughly one-third were black. Vanport is now a symbol of Portland's history of racial segregation formalized as housing policy.

A 1996 Christmas flood was met with Vera's Wall, a flood wall along Waterfront Park erected by volunteers and emergency responders from across the city, and named for former Mayor Vera Katz, who had called for the construction of the wall. Volunteers fashioned a wall out of plastic sheets, sandbags, and plywood scaffolding. The mile-long structure was never put to the test, thankfully, as the Willamette crested at 28.6 feet—never topping the seawall.

The city of San Francisco is aware of the looming dangers of water associated with sea level rise and extreme weather conditions. Vulnerability assessments estimate sea levels rising from six to twelve inches by 2030.

Flooded intersection of SW Third Avenue and SW Washington Street, downtown Portland, 1894.

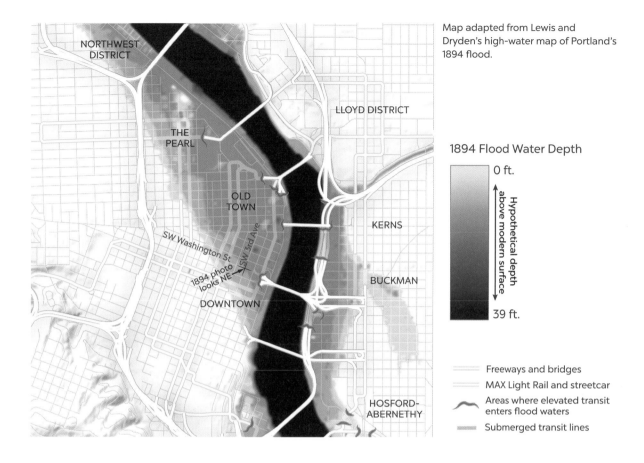

Map adapted from Lewis and Dryden's high-water map of Portland's 1894 flood.

1894 Flood Water Depth

0 ft.

Hypothetical depth above modern surface

39 ft.

— — — Freeways and bridges

———— MAX Light Rail and streetcar

∿∿∿ Areas where elevated transit enters flood waters

▓▓▓▓ Submerged transit lines

Map labels: NORTHWEST DISTRICT · THE PEARL · OLD TOWN · LLOYD DISTRICT · KERNS · BUCKMAN · HOSFORD-ABERNETHY · DOWNTOWN · SW Washington St · SW 3rd Ave · 1894 photo looks NE

THE BIG ONE

What will happen when "the big one" strikes? That question lurks in the shadows of each city's collective unconscious. The phrase speaks of the moment a sudden catastrophic force of nature hits, but a bigger unknown is what would happen next. Long after the rubble is cleared, the ashes swept away, and reorganization begins, the city that was will become something else.

San Francisco is the only Upper Left city where the big one is a firsthand memory, rather than a nebulous threat. In 1989, the Loma Prieta earthquake maimed the eponymous stretch of freeway hunched over a historic bay shore precinct named the Embarcadero. For three decades, the brutalist architecture of the double-decker overpass stood in regrettable contrast with the natural aesthetic of the bay and the historic Ferry Building. After the earthquake, with the physical barrier removed, the city was reunited with these waterfront spaces, a change that has garnered nearly universal praise.

But now, on land decimated by the earthquake, massive—and massively controversial—high-rises are popping up all over the city. The dramatically altered skyline is emblematic of the city's transformation.

The self-image of Upper Left cities is rooted in their connection with surrounding environments—the Bay, the Sound, the Willamette. The past three decades of rebuilding in San Francisco since Loma Prieta offers Seattle and Portland a glimpse into their own post-tragedy futures.

Redevelopment can be a success story or a cautionary tale. Portland and Seattle face an open question following the next big one: Will the inevitable reconstruction minimize connections to the natural environment, or will the recovery be used to rebuild with more consideration of the natural world?

SASQUATCH MEANS BUSINESS

Tales of giant hairy hominids in the forests of the Pacific Northwest predate European settlement. The Salish word for these beasts is *Sasq'ets*. Stories of Sasquatch, and similar creatures (for example, the yeti of the Himalayas), capture the imagination of people the world over. What looks like an obvious prank to some is proof of existence to others.

In 1967, grainy footage emerged of a creature walking along Bluff Creek in Northern California just south of the Oregon border. Though the film is widely believed to be a hoax, the filmmakers have always maintained the veracity of their footage. Known as the Patterson-Gimlin film, the images launched bigfoot into popular culture. It is from this film we get the ubiquitous figure of Sasquatch in full stride.

Some of those who believe in bigfoot because of personal encounters don't call themselves believers—they refer to themselves as knowers. At least one academic researcher, late Washington State anthropologist Grover Krantz, argued passionately for the existence of bigfoot.

Meanwhile, skeptics can't seem to let go of the fact that neither carcass nor skeleton of such a creature has ever been discovered. Or how difficult it would be for entire populations of large mammals to go undetected.

Bigfoot obsession had its Hollywood moment with the 1987 release of the family comedy *Harry and the Hendersons*, about a Seattle family and their live-in Sasquatch. The Academy was so smitten with the creature that the film won an Oscar for Best Makeup. Popular interest seems as strong as ever. *Finding Bigfoot*, a show on Animal Planet, ended in 2018 after nine seasons and one hundred episodes. Bigfoot conventions and festivals are regularly held across the country.

One doesn't have to believe in bigfoot to embrace the creature as a symbol of the wild, a conservation device, or a quirky cultural icon. Bigfoot likenesses are found on everything from mugs to T-shirts to bumper stickers, and are sold everywhere from rural truck stops to urban bookstores.

CATEGORIES OF SASQUATCH-RELATED BUSINESS NAMES

Sasquatch looms large in the Upper Left. Many companies pay homage to the hairy creature in a nod to regional culture—even a certain Seattle-based publisher.

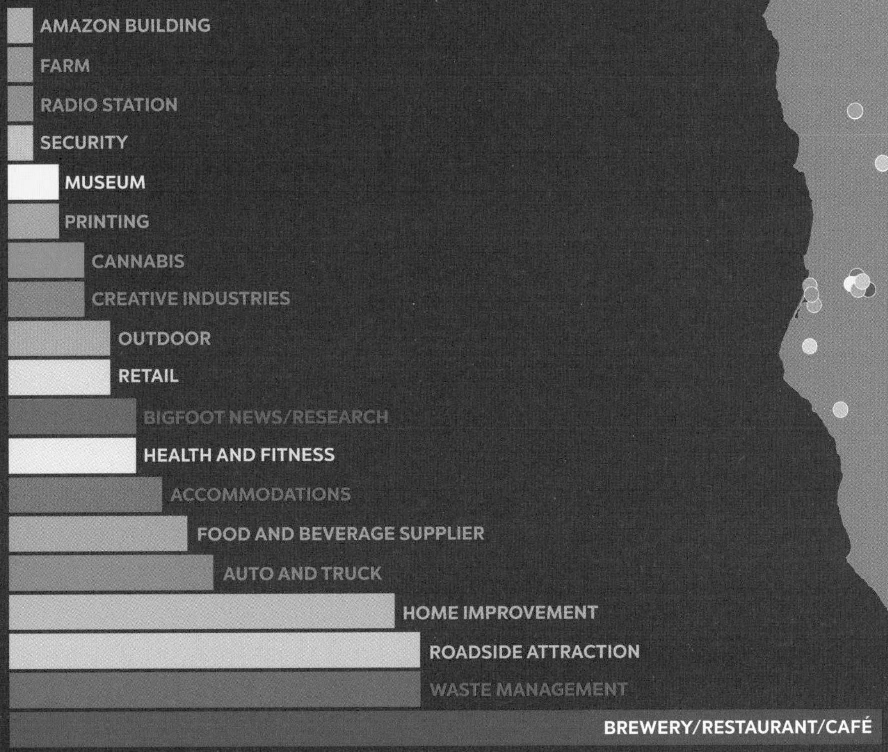

AMAZON BUILDING

FARM

RADIO STATION

SECURITY

MUSEUM

PRINTING

CANNABIS

CREATIVE INDUSTRIES

OUTDOOR

RETAIL

BIGFOOT NEWS/RESEARCH

HEALTH AND FITNESS

ACCOMMODATIONS

FOOD AND BEVERAGE SUPPLIER

AUTO AND TRUCK

HOME IMPROVEMENT

ROADSIDE ATTRACTION

WASTE MANAGEMENT

BREWERY/RESTAURANT/CAFÉ

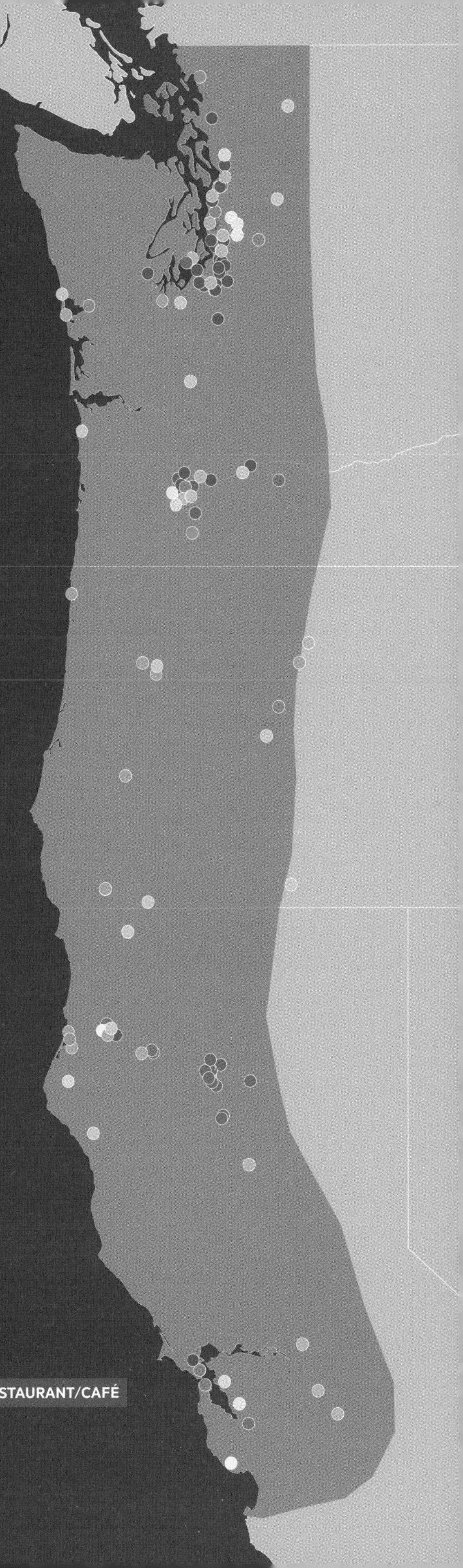

III. SOCIAL RELATIONS

FROM JAPANTOWNS AND JAZZ CLUBS to church bells and ballot boxes, the social life of cities is rich—musically, spiritually, linguistically, politically, and in many other ways. The social life of a city emerges from hundreds of thousands of relationships: local with visitor, artist with patron, dominant culture with subculture.

A city's reputation is often the first thing people learn about a city. Cities in the Upper Left share a reputation for progressivism, though this reputation is sometimes questioned. These cities share significant social challenges—interwoven issues of gentrification, homelessness, and other forms of economic inequity and inequality. People are at odds over how to address these challenges. There is debate over investment priorities, where change should be focused, and at what speed change should happen.

These three cities have reputations as very politically liberal places. This is the popular imagination, which does not match up exactly with reality.

People in San Francisco, Portland, and Seattle are connected not only to each other, but to people all over the world. These cities connect as sister cities, as destinations, and as political symbols. In this chapter, we explore social relations in San Francisco, Portland, and Seattle.

WE ARE FAMILY: SISTER CITIES

As Harper Lee told us, you can choose your friends, but you can't choose your family. Unless you are a city.

Founded in 1956 under President Eisenhower, the US Sister Cities program was established to foster cross-cultural connections and international diplomatic relations after World War II. The goal was to strengthen global relations through cultural, educational, and economic exchange. In 1967, the program evolved into Sister Cities International, a nonprofit organization now working in 140 countries to support international partnerships and citizen diplomacy around the world.

Sister Cities International also facilitates the establishment of Friendship Cities, a less formal relationship for places that aren't quite sure they are ready to be family yet. The map to the right illustrates the sister city network of San Francisco, Portland, and Seattle. On the following pages, we delve deeper into sisterhoods decade by decade.

Khabarovsk
Tashkent
Sapporo
Seoul
Daejeon Kobe
Ulsan Osaka
Chongqing Suzhou Shanghai
Taipei
Haiphong
Kaohsiung
Bangalore
Sihanoukville Manila
Ho Chi Minh City Cebu
Surabaya
Sydney
Christchurch

———— Seattle
———— Portland
———— San Francisco

* Indicated city is just a "friend," not an officially sanctioned sister city

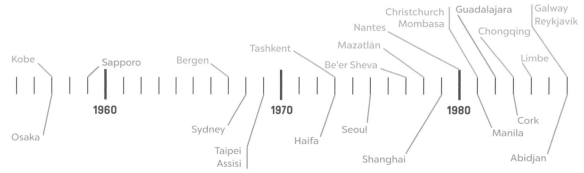

Kobe
Sapporo
Bergen
Tashkent
Christchurch Guadalajara Galway
Nantes Mombasa Reykjavík
Mazatlán Chongqing
Be'er Sheva Limbe

1960
1970
1980

Osaka
Sydney
Seoul
Cork
Manila
Taipei
Assisi
Haifa
Shanghai
Abidjan

Reykjavík

Bergen
Kiel   Gdynia
Galway Utrecht* Kraków
Cork   Paris
Nantes   Zurich Pecs
             Bologna
             Perugia
             Assisi   Thessaloniki
Barcelona

Haifa   Amman
Ashkelon
Be'er Sheva

Mazatlán
Guadalajara

Abidjan   Limbe

Mombasa

Mutare

Kaohsiung
Ashkelon
Ulsan

Kaohsiung
Pecs
Cebu
Perugia

Mutare

Daejeon

Surabaya

Gdynia

Haiphong

Sihanoukville

Bologna

Utrecht*

1990 ── Thessaloniki

Paris

2000

Zurich

2010

Amman
Barcelona

Kiel

Suzhou
Khabarovsk

Ho Chi Minh City

Bangalore
Kraków

SISTER CITIES BY DECADE

SISTERS OF THE 1950s

The late 1950s were about establishing ties with
Japan. San Francisco formed its first sister city
with Osaka in 1957. The same year, Seattle became
sisters with Kobe. Seattle's Kobe Terrace is home
to gifts from the people of Kobe—Mount Fuji
cherry trees and a four-ton, two-hundred-year-old
stone lantern. Portland established its first sister
city relationship with Sapporo in 1959 commemorated at the Friendship Circle in Tom McCall
Waterfront Park.

SISTERS OF THE 1960s

Reflecting Scandinavian ties, Norway's second-
largest city, Bergen, became Seattle's second sister
city in 1967, *fjorging* a series of ongoing cultural
exchanges ever since. San Francisco landed a capital city sibling in 1968 in Sydney, Australia. In 1969,
San Francisco became sisters with the Taiwanese
economic powerhouse Taipei and the tiny hill
town of Assisi in Italy, the home of Saint Francis of
Assisi, for whom San Francisco was named.

SISTERS OF THE 1970s

San Francisco became sister cities with Haifa,
Israel, in 1973 and with the capital of Korea, Seoul,
in 1975. In 1973, Seattle established sisterhood
with Tashkent, Uzbekistan, and became the first
US city to form a Soviet sister city affiliation. The
late 1970s saw Seattle become sister cities with
Be'er Sheva, Israel, and Mazatlán, Mexico's larg-
est Pacific Coast city. During the same time, San
Francisco became sister cities with Shanghai,
China's primary financial and business center.

(top) Terry Schrunk Plaza, Portland, Tai Hu stones gifted from
Suzhou, China, 2014.

(center) Hallidie Plaza, Sister Cities Sign, San Francisco, 2019.

(bottom) Daejeon Park, Seattle, 2019.

SISTERS OF THE 1980s

In 1980, Seattle became sister cities with Nantes, France, and now share a marathon runner exchange. Soon after Seattle and Christchurch, New Zealand, established a similar program. Seattle and Chinese industrial and transportation hub Chongqing formed their sister city association in 1983, commemorated by the Seattle Chinese Garden.

In 1981, San Francisco affiliated with Manila, the economic and cultural center of the Philippines. The same year, Seattle and Mombasa, Kenya's second-largest city, became sister cities, catalyzing educational and trade programs. In 1984, San Francisco became sister cities with tech and research leader Cork, Ireland. In 1986, Abidjan, the economic center of the Côte d'Ivoire, became a sister city of San Francisco.

In 1984, Limbe, Cameroon, and Seattle became sister cities, an association that supports health initiatives for seniors and young girls in Limbe. Seattle formed a sisterhood with Galway in 1986. Seattle established a program with the Icelandic capital port city of Reykjavík, the world's northernmost metropolis. In 1989, the sisterhood between Seattle and the South Korean city of Daejeon began, commemorated by Daejeon Park on Beacon Hill.

Until 1983, Portland's only sister city was Sapporo. However, Portland established six new sister city relationships in the 1980s: Guadalajara, Mexico; Ashkelon, Israel; Kaohsiung, Taiwan; Ulsan, South Korea; Suzhou, China; and Khabarovsk, Russia. Portland's sister cities sometimes share cultural or economic similarities. Khabarovsk is situated at the confluence of two rivers, whose natural resources include salmon and timber. Kaohsiung is another port city and a major trade partner of Oregon. Guadalajara and Portland are both the City of Roses.

SISTERS OF THE 1990s

Seattle's sisterhoods proliferated during the early 1990s, after which the city council passed a moratorium on sisterhoods. New sister cities included harbor city Cebu, Philippines; Hungarian university city Pecs; Italian cultural and economic hub Perugia; Indonesia's second-largest port city, Surabaya; Polish seaport Gdynia; and Vietnam's third-largest city, Haiphong. Seattle also established sisterhood with Taiwanese container-cargo power Kaohsiung, already sister cities with Portland, thereby making Portland and Seattle stepsister cities. In 1999, Cambodian seaport Sihanoukville and Seattle became sister cities.

San Francisco and Greece's second-largest city, Thessaloniki, established their sibling agreement in 1990. In 1995, San Francisco and Ho Chi Minh City became the first-ever US–Vietnam pairing. In 1997, San Francisco giddily became siblings with Paris.

In 1991, Portland established a sister city program with Mutare, Zimbabwe, which promotes cultural exchange and support for schools and medical facilities in Mutare.

SISTERS OF THE 2000s

Portland established a sister city relationship with Bologna, Italy, in 2003, generating one of the most adorable logos in the annals of sister city graphics—a bicycle and a Vespa-style scooter facing each other with a heart between them. In 2012, Portland tepidly established a friendship association in Utrecht, Netherlands.

The first ten years of the 2000s was a prolific time for San Francisco sister city affiliations. Beginning with a relationship with Zurich, Switzerland, San Francisco then became sister cities with Bangalore, known as the Silicon Valley of India; Kraków, Poland; Jordan's capital Amman; and Barcelona, the capital of Catalonia. The most recent sister city relationship in the Upper Left is San Francisco's 2017 siblinghood with Kiel, Germany, a city with a big tech scene.

DEEP PURPLE: VOTING IN THE UPPER LEFT

If you've paid even passing attention to American politics in the last several decades, you've certainly seen a blue and red map of the fifty states trotted out at least once every four years. The large swaths of red and blue obscure the fact that no group of people votes exactly the same way. People who live in the same area are no exception. Ours is a country of individuals, even if the current "winner take all" electoral college system paints an entire state's population with the same red or blue brush.

A finer level of detail reveals a bit more. The divisive election of 2020 provides the perfect opportunity to explore ebbs and flows in spatial partisanship. The national map here breaks the country down into counties or the nearest equivalent (Alaska has boroughs and Louisiana has parishes). We ditched blue and red, instead using purple to represent Democrats and orange to represent Republicans.

Some aspects of conventional wisdom are represented here. It is little surprise, for instance, to see most urban centers as dark purple and many rural counties as dark orange. The pattern of support for Biden on the West Coast and Northeast is clear as is the support for Trump in the plains and mountain states.

What this map does not show is the relative size of population throughout the country. San Francisco, for example, nearly disappears on the map whereas many rural areas with far fewer people take up much more room on the map. The election maps on the next pages provide different views of election data.

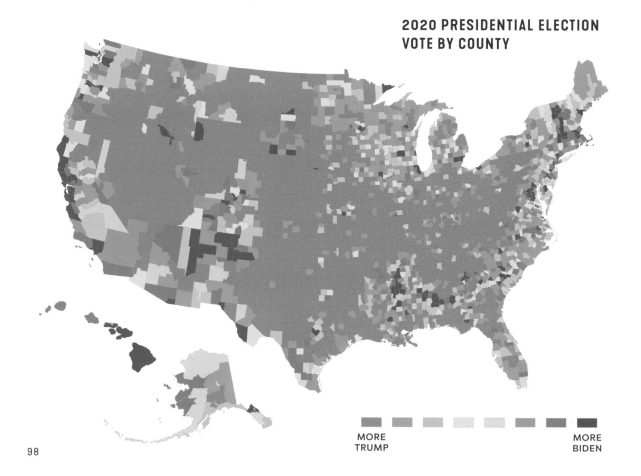

2020 PRESIDENTIAL ELECTION VOTE BY COUNTY

MORE TRUMP · · · · · · · MORE BIDEN

2020 PRESIDENTIAL ELECTION VOTE BY COUNTY: WASHINGTON

2020 PRESIDENTIAL ELECTION VOTE BY COUNTY: OREGON

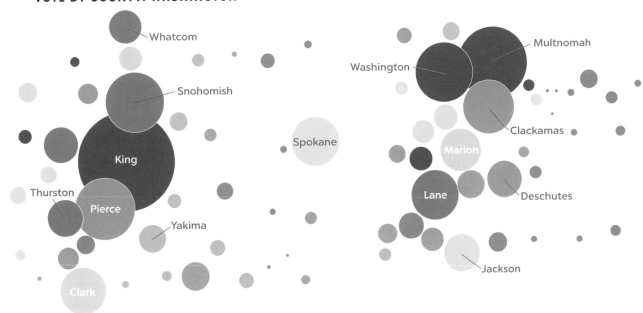

Whatcom

Snohomish

Spokane

King

Thurston

Pierce

Yakima

Clark

Multnomah

Washington

Clackamas

Marion

Lane

Deschutes

Jackson

2020 PRESIDENTIAL ELECTION VOTE BY COUNTY: CALIFORNIA

Solano

Sonoma

Marin

Contra Costa

San Francisco

Alameda

Sacramento

San Mateo

Santa Clara

Los Angeles

San Diego

The three maps on this page are cartograms. Instead of representing relative land area, they map the relative amounts of votes that Biden and Trump received in of California, Oregon, and Washington. This gives us a better view of the partisan landscapes of the Upper Left.

These maps also demonstrate the relative sizes of populations within each state. However, the scale of the map is particular to each state so these population circles cannot be compared from state to state. If we made the scale the same for all three states, the need to fit the California circles on the page would render many circles in Oregon and Washington as tiny dots. Like much of the country, many parts of the Upper Left had urban support for Biden and rural support for Trump. However, each state has electoral surprises that provide texture to the West Coast's political geography.

California has a predictably sharp divide between the urban coast and rural areas in the interior and northern regions of the state. Although strong regional divergences do exist, the San Francisco metro area was predictably deep purple. The graduated circles also reveal just how much larger the population of Southern California is compared to that of the San Francisco Bay area.

Oregon's political divides seem more clearly separated. The urban centers in and around the Willamette Valley are strongly purple, particularly Multnomah County, home to Portland. A few regions along the coast broke purple whereas the eastern two-thirds of the state, outside of Bend in Deschutes County, are staunchly orange. The graduated symbols illustrate how heavily populated the Willamette Valley is compared to the eastern part of the state.

Washington's rural areas are a relative patchwork of varying shades of purple and orange compared to Oregon's. They tell a story of a more varied electorate, as well as a state with a less-centralized population. The deepest pockets of dark orange are generally situated to the east of Yakima and south of I-90, with the exception of the light orange found in Spokane. Breaking the

pattern of orange in southeastern Washington is Whitman County, home to Washington State University. King County, where Seattle is located, is deep purple.

Voters in every precinct of San Francisco, Portland, and Seattle chose Biden over Trump. By a lot. In the areas of the city that Trump got the most votes, Biden still generally got two or three times as many. These maps provide some evidence for the oft-heard claims that the three cities are liberal bubbles. Still, there are geographies to Biden's support in each city as we have illustrated here.

In Seattle, the absolute deepest support for Biden included large parts of the Central District and Capitol Hill, although that dominance was relatively less pronounced in Montlake. In Portland, Biden's support was strongest in the inner northeast and southeast parts of the city. East Portland showed comparatively less enthusiastic support for Biden—who still handily won these areas. The center of San Francisco was the absolute core of Biden's support. This included large parts of the Haight, Noe Valley, and the Castro. Biden's support was slightly less extreme in the south of the city in places such as Portola, Visitacion Valley, the Excelsior, Lakeshore, and a chunk of Outer Sunset.

PERCENTAGE OF TOTAL STATE VOTES CAST BY CITY AND METRO REGION, 2020

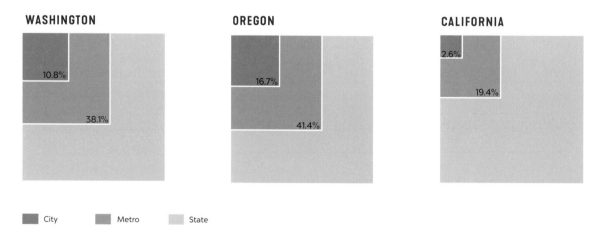

WASHINGTON
10.8%
38.1%

OREGON
16.7%
41.4%

CALIFORNIA
2.6%
19.4%

■ City   ■ Metro   ■ State

SAN FRANCISCO

SEATTLE

PORTLAND

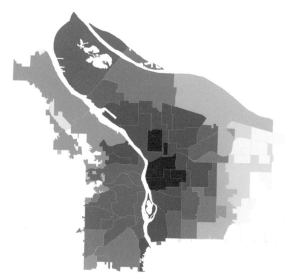

RATIO OF BIDEN TO TRUMP VOTERS

■ More than 20 to 1	■ More than 3 to 1
■ More than 10 to 1	■ More than 2.5 to 1
■ More than 7.5 to 1	□ More than 2 to 1
■ More than 5 to 1	□ Less than 2 to 1
■ More than 4 to 1	

ALL POLITICS ARE LOCAL

The graphs and maps on these pages examine voter turnout in local elections. Top to bottom shows voter turnout. The dots at the top of the graph represent areas with higher voter turnout and those at the bottom, lower voter turnout. The distance from the centerline (either right or left) shows relative population of areas. The closer a dot is to the centerline, the smaller a population it has.

Here we display San Francisco and Seattle on the same graph with two corresponding inset maps. The San Francisco 2015 mayoral election is shown alongside Seattle's 2017 mayoral election. For this particular pair of elections, San Francisco has seven districts where turnout exceeds 50 percent and no districts that fall below 40 percent. There was no Seattle district that exceeded 50 percent; two fell below 40 percent turnout. In San Francisco, the smallest district had the highest turnout, whereas

SAN FRANCISCO DISTRICTS

SEATTLE DISTRICTS

SAN FRANCISCO
2015 mayoral election

SEATTLE
2017 mayoral election

Increasing voter turnout

Increasing population in district

Seattle's two smallest districts had the smallest and third-smallest turnouts in the city.

This graph of Portland shows the vote for city council from 2016. Compared to the graph of San Francisco and Seattle, there are many more precincts here, so we chose to look at Portland by itself. We broke the city into six quadrants—although not exactly the city-designated quadrants explored in the introduction.

In Portland, the east part of the city has the lowest turnout overall, although the absolute lowest turnout was in a district in Southwest Portland.

Portland's turnout is noticeably higher than in San Francisco and Seattle. This has much to do with the fact that all voting in the state of Oregon is done by mail, which gives the state one of the highest voter turnout rates in the country.

Even if all of these voter turnout percentages seem low, they are actually pretty good for local elections when compared to much of the country. For example, in a study of local election voter turnout in fifty US cities, researchers at Portland State University found that turnout was under 15 percent in ten of the country's thirty largest cities—three of those cities had turnouts in the single digits. The same study put Portland's voter turnout at the top of the fifty cities with nearly 60 percent. Seattle was third, having a turnout of about 45 percent, and San Francisco sixth with 32 percent turnout.

PORTLAND "QUADRANTS"

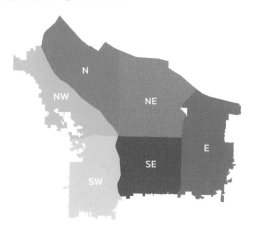

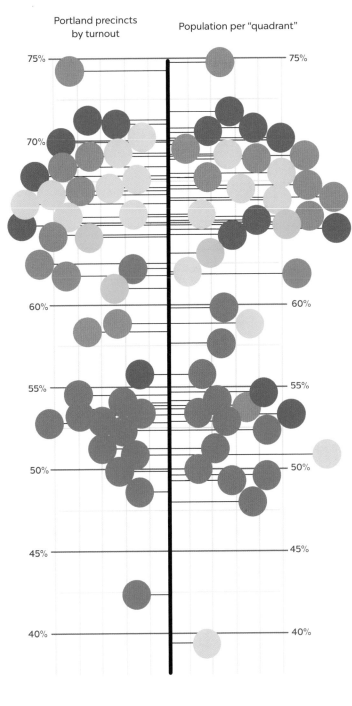

PORTLAND

2016 city council election

Portland precincts by turnout

Population per "quadrant"

103

BLUE NOTES: LOST JAZZ CLUBS

When it comes to music, San Francisco, Portland, and Seattle are probably much better known for rock stars than for jazz artists. Yet each city has a rich jazz history of its own. Three streets figure prominently in these histories: Fillmore Street in San Francisco, Williams Avenue in Portland, and Jackson Street in Seattle. The stories of jazz clubs on these streets illustrate how economics, popular culture, and racial politics combine and change places.

Fillmore, Williams, and Jackson were not the only streets where jazz thrived in these cities. However, each hosted a scene at a time when jazz was at the height of its national popularity and at a time when ethnic and racial segregation was in full effect. Many jazz clubs were black-owned at a time when nonwhites were not permitted in white-owned clubs. These streets anchored the heart of jazz scenes in neighborhoods that were central to black economic and cultural life and were some of the only areas in these cities where black people could live.

During the 1950s and 1960s, redlining, blockbusting, and urban renewal took an enormous toll on these neighborhoods and their jazz clubs. In addition to these racial and urban politics, jazz was becoming less popular. While many jazz clubs have disappeared, the music has not.

Fillmore Street, San Francisco
Heart of the San Francisco jazz scene, 1930s–1950s

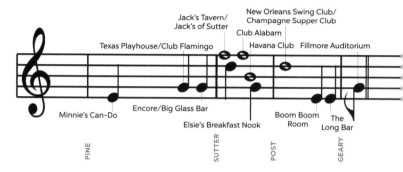

Williams Avenue, Portland
Heart of the Portland jazz scene, 1930s–1950s

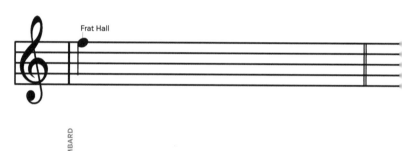

Jackson Street, Seattle
Heart of the Seattle jazz scene, late 1910s–1950s

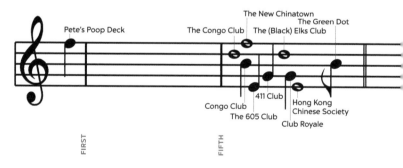

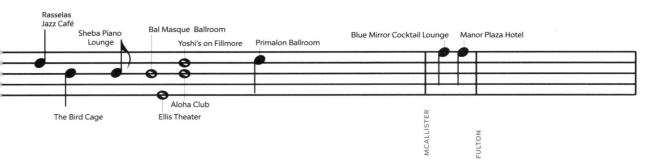

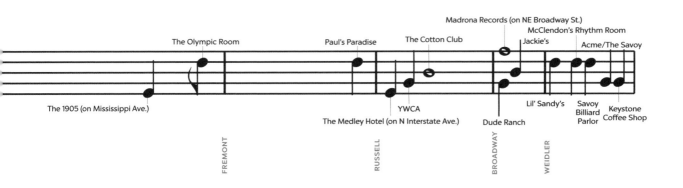

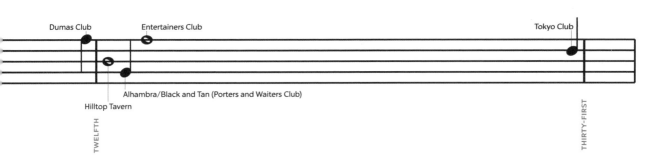

JAZZ CLUBS OF JACKSON STREET, SEATTLE

Jackson Street was the heart of jazz in Seattle from the late 1910s to the 1950s. At a time of extreme housing discrimination against nonwhite people, Jackson Street was a place where black people could own clubs, work as musicians, and enjoy music.

Jackson Street in the early twentieth century had a diverse mix of people, primarily Jewish and Japanese immigrants, as well as many black, Italian, and Chinese residents. Japanese businesses were widespread around this area until World War II and the US government's internment of people of Japanese ancestry. Jackson Street connected Japantown and Chinatown to the predominantly black neighborhood further east.

In the 1940s, wartime industries attracted many black workers and families from the rural south to Seattle (similarly to San Francisco and Portland). From 1940 to 1950, the number of black people living in Seattle quadrupled.

In the early years, E. Russell "Noodles" Smith heavily influenced the development of jazz in Seattle. He opened clubs along Jackson Street and beyond, including the Entertainers Club, the Alhambra (which became the Black and Tan), and the Hilltop Tavern. In the International District, he

owned both the Golden West Hotel and the Coast Hotel, which hosted many nationally known jazz musicians touring through the city.

Several artists from the jazz scene on Jackson Street would go on to achieve national and international success. Ernestine Anderson went from the clubs of Jackson Street to fifteen months of touring with Lionel Hampton to the cover of *Time* magazine in 1958. Ray Charles and Quincy Jones met in Seattle as teenagers, the former helping educate his younger friend about jazz composition and arrangement. While Charles spent only two years in Seattle, they were formative ones, what he called his bar mitzvah.

(top) Ernestine Anderson sign on Jackson Street, Seattle, 2019.

(left) Group around table at the Black and Tan, Seattle, circa 1947.

The Jackson Street jazz era began to fade in the late 1940s and early 1950s. Police raids relentlessly targeted after-hours clubs and speakeasies, the lifeblood of the scene. A 1949 Washington State law that legalized selling liquor by the drink further catalyzed a crackdown on illegal "bottle clubs." City authorities, supported by Teamsters looking to corner the market for their restaurant, hotel, and bar workers, shut down club after club.

In the 1950s, as Jackson Street's jazz scene was being shuttered, a new scene for jazz developed downtown in cocktail lounges largely under union influence. The liquor downtown joints served was legal, even if the payoff-enabled gambling was not. Initially, black musicians were not permitted to play in downtown clubs.

JAZZ CLUBS OF THE FILLMORE, SAN FRANCISCO

By the early twentieth century, the Fillmore had become the most racially and ethnically diverse neighborhood in the city, with Japanese, Filipino, Mexican, Russian, black, and white people living in the same area. This diversity stemmed in part from the barring of nonwhite groups of people from living elsewhere in the city. The Fillmore emerged as the heart of San Francisco's jazz scene from the 1930s into the 1950s.

World War II brought an influx of black migrants to the Bay Area; these migrants often worked in military industries, particularly shipyards. Due to discriminatory housing practices, the Fillmore and nearby Western Addition were about the only areas in San Francisco where African Americans could find housing.

Although North Beach and Divisadero had concentrations of jazz clubs, those on Fillmore became renowned. Jack's Tavern, or Jack's of Sutter, the first club in the area associated with the Fillmore jazz scene, opened in 1933. Jack's was also the first African American club on Fillmore.

An anchor of the Fillmore jazz scene for a time was Jimbo's Bop City, located on Post Street two blocks from Fillmore Street. First a Japanese-owned drugstore, the building became the short-lived Vout City before becoming Jimbo's Waffle Shop, then Bop City, and later Minnie's Can-Do Club, which opened in 1940 and remained in the same location until 1974.

View of Fillmore Street from Bush, looking south, Club Flamingo on left, 1965.

As with Bop City, many clubs were renamed, some more than once, over a period of just a few years. The Club Alabam, which opened in 1935, became Club Iroquois in 1947, Club Sullivan in 1948, and Club Alabam again in 1950 (until its closing in 1953). The New Orleans Swing Club of the 1940s became the Champagne Supper Club in the 1950s. Elsie's Breakfast Nook became Café Society and then closed six years later in the late 1940s.

The era ended when the city's Redevelopment Agency declared the vibrant area blighted and slated it for redevelopment. This meant the leveling of all structures on entire blocks to make way for public housing and other new development. Along with jazz clubs, many other businesses, homes, schools, and churches were destroyed.

Evading the fate of many other buildings that hosted historic jazz clubs, the Victorian house that hosted Vout City and Bop City was moved a short distance to 1712 Fillmore Street, housing Marcus Books, the oldest African American bookstore in the country, until they were evicted in 2014. Jack's moved to 1601 Fillmore in 1988, becoming John Lee Hooker's Boom Boom Room in the 1990s. The club still operates today as the Boom Boom Room.

JAZZ CLUBS OF WILLIAMS AVENUE, PORTLAND

World War II catalyzed the migration of black workers primarily from the rural South to Portland for employment in shipbuilding and other military industries. A hastily constructed housing project called Vanport became home to most of these black migrant workers and their families. After Vanport flooded in 1948, North Portland became the new center of black economic and cultural life, and Williams Avenue, the center of the jazz scene in Portland.

Initially constructed as a hazelnut ice-cream factory and later used as a speakeasy, 240 North Broadway became the Dude Ranch, a black-owned jazz club. Pat Patterson, the first African American to play basketball at the University of Oregon, and friend Sherman Pickett operated the Dude Ranch, the most popular club of its time. Bandleader Al Pierre opened the Frat Hall in 1930, another of the prominent black-owned businesses in the neighborhood.

In 1943, the Acme Club opened on 1500 North Williams and Ed Slaughter's Savoy Billiard Parlor operated on the floor below. The Acme Club served as a hot spot for jazz even after its name change to the Savoy in 1947 and McClendon's Rhythm Room in 1949.

By 1958, the jazz era on Williams Avenue had begun to fade. The area was targeted for urban development and highway construction, fracturing the neighborhood. Nationally touring luminaries visited Portland's jazz scene less and less. As air travel became increasingly viable for musicians, Portland was no longer a sensible stop between San Francisco and Seattle, but a flyover city.

Duke Ellington, McElroy's Ballroom, before heading to McClendon's for his birthday celebration, Portland, 1953.

McElroy's Spanish Ballroom was located apart from Williams Avenue in downtown Portland at SW Fifth and Main, where the Portland Building now stands. The ballroom's white owner, Cole "Pop" McElroy, opened McElroy's to black people on designated nights, something atypical for white owners in the 1930s. It wasn't until 1949 that black and white audiences could finally dance together at McElroy's.

A new wave of jazz clubs opened and operated in the 1970s and 1980s, but several forces worked against the viability of jazz clubs in Portland. For example, many different venues, such as the Portland Art Museum, began to feature acts that previously would have played at jazz clubs. Also, jazz festivals may have watered down the availability of larger acts outside of festival season. In 1984, Portland had half as many venues featuring live jazz as it did in 1980. Jazz didn't disappear—it became more concentrated.

San Francisco jazz artists Kim Nalley (standing) and Tammy Hall (at piano), Fillmore Jazz Festival, San Francisco, 2017.

JAZZ TODAY

Today, there are only a handful of clubs in each city solely dedicated to jazz. Despite the few number of jazz-only clubs, each city has a jazz scene filled with accomplished and passionate musicians who will play in kombucha bars if it means getting to play live.

With the 2019 closure of Tula's Restaurant and Jazz Club, the easiest place to catch jazz in Seattle today is probably downtown's Dimitriou's Jazz Alley, which features both local and nationally touring musicians. The Earshot Jazz Festival provides Seattleites with over a month of additional opportunities to hear live jazz. Ballard also hosts an annual jazz festival at the end of May.

In 2013, the San Francisco Jazz Center opened in Hayes Valley. The venue, which includes a seven-hundred-seat concert hall, is run by the nonprofit organization that runs the annual San Francisco Jazz Festival. Every July, a separate event called the Fillmore Jazz Festival is held on twelve blocks of Fillmore Street closed to traffic. It is billed as the largest free jazz festival on the West Coast.

Despite the decline in clubs, jazz still has a presence in Portland. KMHD, an FM station dedicated exclusively to jazz, has been broadcasting since 1984. KOPB and KBOO also dedicate significant airtime to jazz programming. The PDX Jazz Festival began in 2004 and is still held every February. Cathedral Park and the Montavilla neighborhood also host other annual jazz festivals. The 1905 (fifteen blocks west of Williams Avenue) features live jazz every night of the week.

ENCAMPMENTS

San Francisco, Portland, and Seattle each have a homelessness crisis. And each city has admitted failure in their ability to deal with the situation. Portland and Seattle (and King County) declared homelessness a state of emergency in 2015. San Francisco did the same the following year. Doing this did not allow the cities to access any emergency funding as cities sometimes do after natural disasters. Rather, the declarations highlight the ongoing debates about how to provide for the homeless as tent cities seem to be more and more common.

The United States Department of Housing and Urban Development conducts point-in-time counts to make rough estimates of homeless populations. Point-in-time counts are taken annually across the country on the last ten days of January.

However, local levels of government, nonprofit organizations, and media outlets are also involved in collecting data about homelessness. Cities and counties conduct their own point-in-time counts as well. This count is conducted by volunteers. Although it misses many people, it is the most extensive spatial data we've encountered on homelessness.

Controversies even surround how to count and map the homeless and whether it should be done at all. In 2017, a local news station in Seattle crowdsourced data on homeless encampments and posted it on their website. The encampments reported were all clearly visible from the sidewalk, not hidden. Regardless, strong criticism on social media followed.

The station, criticized for endangering homeless people by showing their locations, responded by suggesting the city should make such data available. The station did replace the map of specific locations with another that showed generalized areas of encampments. This speaks to larger debates about mapping vulnerable populations and whether that does more help or harm to those groups of people.

San Francisco and Portland both have a city-run program through which people can report encampments. Portland's PDX Reporter app takes citizen data on campsites, in addition to another dozen complaints that include graffiti, potholes, and abandoned cars. San Francisco's SF311 app takes reports on a similar list of citizen concerns. For both cities, the data is publicly and readily available for the asking.

Mapped here are the 2019 point-in-time count for San Francisco, the 2018 One Point of Contact Campsite reports for Portland, and Seattle's "emphasis areas" for encampment removal in 2019. It's important to note that crowdsourced data like we show for Portland can be misleading. Many people may report a given campsite, so it could present as dozens of dots instead of one.

The largest concentration of reports for encampments in San Francisco is an area that includes downtown, South of Market, the Tenderloin, and part of the Mission. Reports are fewer on the west side of the city. In Portland, the most encampment reports are made for downtown, the Southeast Industrial District, the Lloyd District, and parts of East Portland by I-205 and the light-rail lines. In Seattle, the emphasis areas include the waterfront near downtown, the area around Lumen Field and T-Mobile Park, the Mercer Corridor, and the East Duwamish Greenbelt.

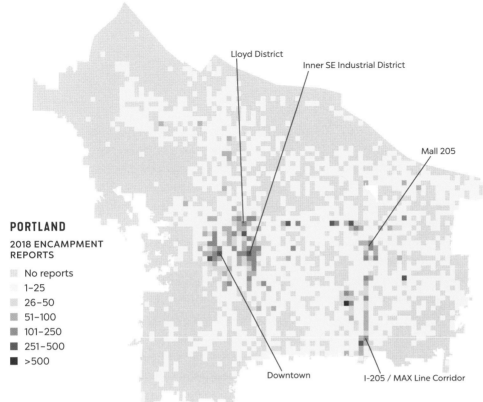

PORTLAND

2018 ENCAMPMENT REPORTS

- No reports
- 1–25
- 26–50
- 51–100
- 101–250
- 251–500
- >500

Lloyd District

Inner SE Industrial District

Mall 205

Downtown

I-205 / MAX Line Corridor

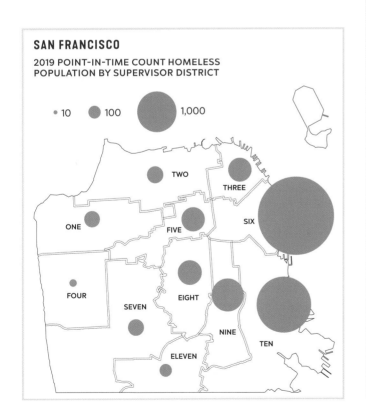

SAN FRANCISCO

2019 POINT-IN-TIME COUNT HOMELESS POPULATION BY SUPERVISOR DISTRICT

- 10
- 100
- 1,000

ONE
TWO
THREE
FOUR
FIVE
SIX
SEVEN
EIGHT
NINE
TEN
ELEVEN

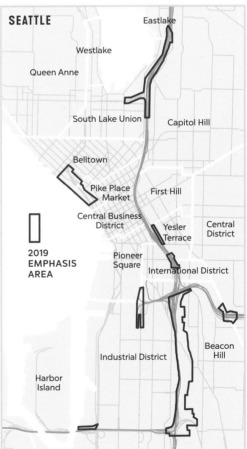

SEATTLE

Eastlake
Westlake
Queen Anne
South Lake Union
Capitol Hill
Belltown
Pike Place Market
First Hill
Central Business District
Yesler Terrace
Central District
Pioneer Square
International District
Harbor Island
Industrial District
Beacon Hill

2019 EMPHASIS AREA

THERE GOES THE GAYBORHOOD

The "fight for freedom" is no misnomer. In 1960s New York City, it was common for police to raid gay bars and arrest patrons. On June 28, 1969, police in New York City raided the Stonewall Inn in Greenwich Village, but as arrests began to happen a crowd formed around the club and refused to disperse. The raid and ensuing riot sparked six days of demonstrations and citizen conflict with police. It also helped ignite the Gay Liberation movement and is widely recognized as a watershed moment in LGBTQ civil rights.

The Stonewall Inn was a place to drink and dance, a dive bar enshrined for civil rights veneration. But what becomes of its kin, other dives with windows painted over and bottles sporting premium labels filled with cheaper liquor? What happened to its social descendants, the bars and dance clubs?

Like their millennial counterparts, bars increasingly shirk binary labels. In the 1970s, to increase cachet and business, blue-collar watering holes rebranded themselves as gay bars. Today, they rebrand for the same reason, not finding it good business practice to limit clientele with a diminishing neighborhood demographic.

The Castro in San Francisco is perhaps one of the best-known gay neighborhoods in the country. In Seattle the corresponding neighborhood is Capitol Hill. Portland's gay neighborhood by contrast is a slim triangle of a few blocks downtown.

Fifty years after Stonewall, we focused on the opening and closing of bars and clubs that identify as LGBTQ focused. The dates on the following maps may read like tombstones to some, but these are not cemeteries. As a gayborhoods change and migrate, they continue to provide vital and distinct LGBTQ spaces in the city.

Now, where do you want to get a drink?

Crosswalk in the Castro, San Francisco, 2017.

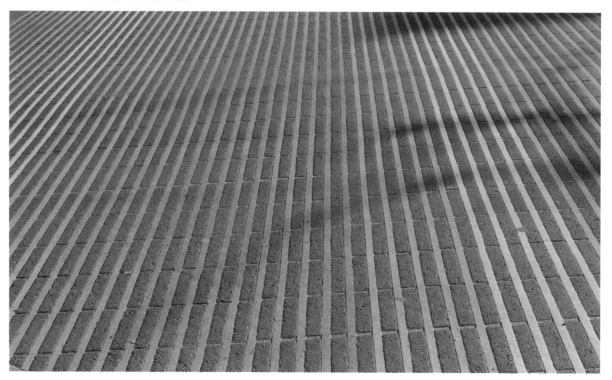

SEATTLE

In 1967, two years before Stonewall, the Dorian Society, Seattle's first gay organization, formed. Its mission was to provide services for the gay community and promote acceptance of homosexuality in mainstream society. Seattle held its first Gay Pride Week in 1977.

Seattle is known as the Queen City for good reason. In 2010 it was reported to be one of the gayest cities in America. When a remodel of Pioneer Square in the 1970s displaced gay businesses, Capitol Hill was more known for car dealerships. During the 1970s when gay men were socially transforming Capitol Hill, women created lesbian feminist spaces in the University District.

As in many other cities, lesbian social spaces coalesced apart from the gayborhood.

By the late 1980s, Capitol Hill was firmly established and known as a gayborhood, with its heyday running through the early 2000s. Property values increased, hip areas attracted new residents, and now Capitol Hill is home to fewer LGBTQ people than in past decades.

Still, a handful of Capitol Hill bars have stood the test of time. The Crescent Lounge, currently a favorite spot for karaoke, is thought to have been gay-owned and operated in the 1960s. The Seattle Eagle, Madison Pub, and Wildrose opened in the 1980s and continue to operate today.

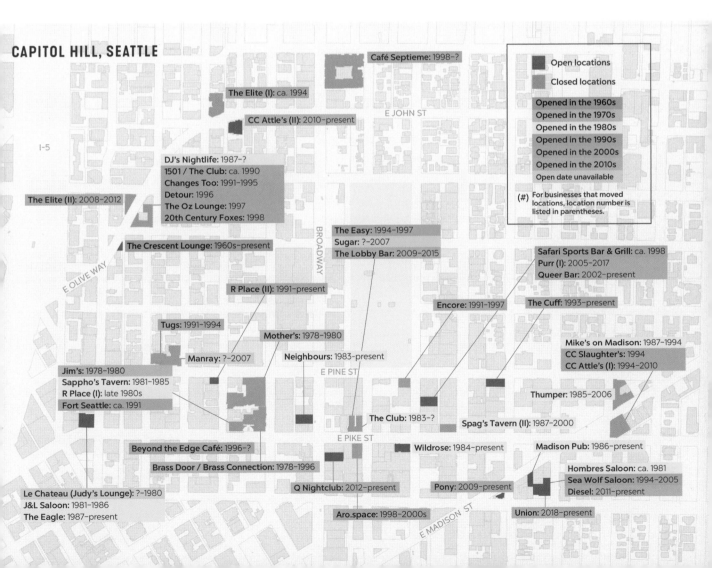

CAPITOL HILL, SEATTLE

Café Septieme: 1998–?

The Elite (I): ca. 1994

CC Attle's (II): 2010–present

E JOHN ST

Legend:
- Open locations
- Closed locations
- Opened in the 1960s
- Opened in the 1970s
- Opened in the 1980s
- Opened in the 1990s
- Opened in the 2000s
- Opened in the 2010s
- Open date unavailable
- (#) For businesses that moved locations, location number is listed in parentheses.

DJ's Nightlife: 1987–?
1501 / The Club: ca. 1990
Changes Too: 1991–1995
Detour: 1996
The Oz Lounge: 1997
20th Century Foxes: 1998

The Elite (II): 2008–2012

I-5

BROADWAY

The Easy: 1994–1997
Sugar: ?–2007
The Lobby Bar: 2009–2015

Safari Sports Bar & Grill: ca. 1998
Purr (I): 2005–2017
Queer Bar: 2002–present

The Crescent Lounge: 1960s–present

E OLIVE WAY

R Place (II): 1991–present

Encore: 1991–1997

The Cuff: 1993–present

Tugs: 1991–1994

Mother's: 1978–1980

Mike's on Madison: 1987–1994
CC Slaughter's: 1994
CC Attle's (I): 1994–2010

Manray: ?–2007

Neighbours: 1983–present

E PINE ST

Jim's: 1978–1980
Sappho's Tavern: 1981–1985
R Place (I): late 1980s
Fort Seattle: ca. 1991

Thumper: 1985–2006

The Club: 1983–?

Spag's Tavern (II): 1987–2000

E PIKE ST

Beyond the Edge Café: 1996–?

Wildrose: 1984–present

Madison Pub: 1986–present

Brass Door / Brass Connection: 1978–1996

Hombres Saloon: ca. 1981
Sea Wolf Saloon: 1994–2005
Diesel: 2011–present

Le Chateau (Judy's Lounge): ?–1980
J&L Saloon: 1981–1986
The Eagle: 1987–present

Q Nightclub: 2012–present

Pony: 2009–present

Union: 2018–present

Aro.space: 1998–2000s

E MADISON ST

SAN FRANCISCO

San Francisco's Stonewall moment occurred three years prior, in 1966, when a drag queen at Gene Compton's Cafeteria in the Tenderloin threw coffee in the face of a policeman trying to arrest her. Even before that, *Life* magazine called San Francisco "the gayest city in America." In addition to the Tenderloin, the area around Polk Street attracted affluent midcentury gay people, while South of Market catered to adult tastes and leather styles. So, in the 1970s, the Castro became a mecca to which gay Midwesterners and Westerners could make pilgrimage.

In 2019, the city of San Francisco officially recognized the Castro as a historic and cultural district worth preserving. The area's only lesbian bar had moved to Hayes Valley. The last remaining Latino club was closed in 2014.

Perhaps gay tourism draws the most people to "the capital of liberated gay America"; this may ultimately save the Castro. Drinks still flow on the site of its first gay bar. Indeed, gay bars rise like phoenixes.

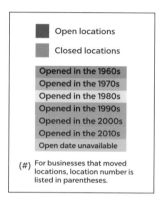

■ Open locations

■ Closed locations

Opened in the 1960s
Opened in the 1970s
Opened in the 1980s
Opened in the 1990s
Opened in the 2000s
Opened in the 2010s
Open date unavailable

(#) For businesses that moved locations, location number is listed in parentheses.

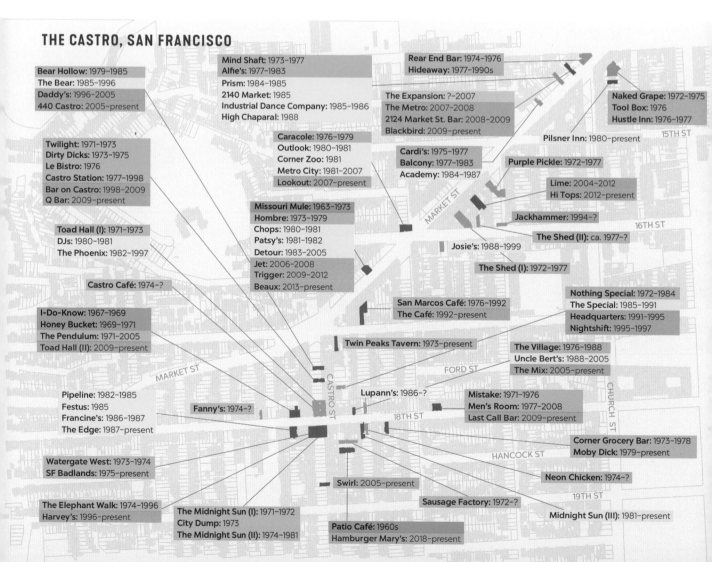

THE CASTRO, SAN FRANCISCO

Bear Hollow: 1979–1985
The Bear: 1985–1996
Daddy's: 1996–2005
440 Castro: 2005–present

Mind Shaft: 1973–1977
Alfie's: 1977–1983
Prism: 1984–1985
2140 Market: 1985
Industrial Dance Company: 1985–1986
High Chaparal: 1988

Rear End Bar: 1974–1976
Hideaway: 1977–1990s

The Expansion: ?–2007
The Metro: 2007–2008
2124 Market St. Bar: 2008–2009
Blackbird: 2009–present

Naked Grape: 1972–1975
Tool Box: 1976
Hustle Inn: 1976–1977

15TH ST

Pilsner Inn: 1980–present

Twilight: 1971–1973
Dirty Dicks: 1973–1975
Le Bistro: 1976
Castro Station: 1977–1998
Bar on Castro: 1998–2009
Q Bar: 2009–present

Caracole: 1976–1979
Outlook: 1980–1981
Corner Zoo: 1981
Metro City: 1981–2007
Lookout: 2007–present

Cardi's: 1975–1977
Balcony: 1977–1983
Academy: 1984–1987

Purple Pickle: 1972–1977

Lime: 2004–2012
Hi Tops: 2012–present

Jackhammer: 1994–?

16TH ST

Toad Hall (I): 1971–1973
DJs: 1980–1981
The Phoenix: 1982–1997

Missouri Mule: 1963–1973
Hombre: 1973–1979
Chops: 1980–1981
Patsy's: 1981–1982
Detour: 1983–2005
Jet: 2006–2008
Trigger: 2009–2012
Beaux: 2013–present

The Shed (II): ca. 1977–?

Josie's: 1988–1999

The Shed (I): 1972–1977

Castro Café: 1974–?

San Marcos Café: 1976–1992
The Café: 1992–present

Nothing Special: 1972–1984
The Special: 1985–1991
Headquarters: 1991–1995
Nightshift: 1995–1997

I-Do-Know: 1967–1969
Honey Bucket: 1969–1971
The Pendulum: 1971–2005
Toad Hall (II): 2009–present

Twin Peaks Tavern: 1973–present

The Village: 1976–1988
Uncle Bert's: 1988–2005
The Mix: 2005–present

FORD ST

MARKET ST

CASTRO ST

CHURCH ST

Pipeline: 1982–1985
Festus: 1985
Francine's: 1986–1987
The Edge: 1987–present

Fanny's: 1974–?

Lupann's: 1986–?

Mistake: 1971–1976
Men's Room: 1977–2008
Last Call Bar: 2009–present

18TH ST

Watergate West: 1973–1974
SF Badlands: 1975–present

Corner Grocery Bar: 1973–1978
Moby Dick: 1979–present

HANCOCK ST

Neon Chicken: 1974–?

Swirl: 2005–present

Sausage Factory: 1972–?

19TH ST

The Elephant Walk: 1974–1996
Harvey's: 1996–present

The Midnight Sun (I): 1971–1972
City Dump: 1973
The Midnight Sun (II): 1974–1981

Patio Café: 1960s
Hamburger Mary's: 2018–present

Midnight Sun (III): 1981–present

DOWNTOWN, PORTLAND

Mocambo/The Pied Piper/
Kuchina Lounge: 1960–1976
Riddles: 1976–1977
The Bushes: late 1970s
Stark St. Station: 1982
Flossie's (II): 1982–1983
Silverado (I): 1983–2008

The Royale: 2014
Stag: 2015–present

Dirty Duck: 1984–2009

Demas Tavern: 1968–1973
Darcelle XV: 1974–present

The Cell / JR's West: 1981–1984

Club Northwest: 1968–1973

Fox and Hounds: 1990–2017

Embers (II): 1981–2018

CC Slaughters (II) / Rainbow Room: 1997–present

Somebody's Place (II): 1983–1985

Hobo's: 1978–present

Casey's (I): 2014–2018
Silverado (III): 2019–present

Hamburger Marys (IV): 2010–2013

Aunt Fanny's: 1975

City Nightclub (II): 1989–1997

Boxxes: ?–2012

The Brig: ?– 2012

Sportsman's Inn: ca. 1972

Rising Moon: 1978–1985

PDX Eagle: ?–2008

Red Cap: 1987–2012

Teasers: ca. 1985

Panorama: ?–2001

Annex: 1965–1971
Family Zoo: 1971–1985

316: 1974

214 Tavern: ca. 1985

Three Sisters (II): 1963–2005
Scandals (II): 2006–present

Hamburger Mary's (II): ca. 1978

Mildred's Palace (II): early 1980s
Metropolis: ca. 1982

The Rage: ?–1998
Escape: 2003–2017

City Nightclub (I): 1983–1989

The Riptide: 1965–1969
Roman's Riptide: 1969–1973

Invasion Lounge: 2009–2010
Casey's (II): 2010–2014

Wilde Oscar's: 1976–1983
Silverado (II): 2008–2018

Hunt Pub: 1983

CC Slaughters (I): 1982–1997

Timber Topper: 1970–1974
Axe Handle: 1974
The Alley: 1970s

Scandals (I): 1985–2005

Zorba the Greek's: 1971–1976

The Watercrest: ca. 1975

Ritz Disco: mid-1970s
Flight 181: 1976–?

The Focal Point: 1971–1974
Fiddlers Three: 1974–1979
Somebody's Place (I): 1979–1983

Grand Oasis Tavern: 1980s

The Other Inn: 1964–1982

Mildred's Palace (I): 1977–ca. 1979

Castaways Lounge: 1972–?

Dahl & Penne Tavern: 1960s–1983

Roman's Tavern: 1969–1971
The Rafter's / Embers (I): 1971–1981

SW HARVEY MILK ST
SW 12TH AVE
SW PARK AVE
SW ALDER ST
W BURNSIDE ST
NW DAVIS ST
NW 3RD AVE
I-405

PORTLAND

City codes born of moral outrage following the Vice Clique scandal of 1912 influenced policy toward the gay community for the next half century.

In 1970, a singles ad ran in a Portland weekly paper that read, "gay and lonely seeking same." For those who answered the ad, a reply came proposing a Gay Liberation Front much like New York did following Stonewall. Since then, Portland has earned its liberal reputation.

In 2018 Portland renamed a portion of Southwest Stark Street, christening it SW Harvey Milk Street. But the renaming of the main gay drag is as much commemoration as it is memorial. In a game of musical chairs, LGBTQ clubs once lining the street have relocated, often to fronts vacated by sister clubs. For some clubs, the lease ran out and the game ended.

Yet in Old Town, the Empress Emeritus of Drag, Darcelle XV, still entertains at a show that premiered before Stonewall, a Guinness World Record. And Portland innovates. In a city well known for strip clubs, 2019 saw the beginning of one of the nation's first regular transgender strip shows.

UPRISING

George Floyd was killed by Minneapolis police on May 25, 2020—Memorial Day. While Floyd was handcuffed and face down on the ground, a white police officer knelt on his neck for over eight minutes. The murder was recorded by a bystander and soon broadcast across the world. By that weekend, protests erupted across the United States. After more than two months of lockdown from COVID-19, people took to the streets. Demonstrations in San Francisco, Portland, and Seattle began the last weekend in May.

From the time protests began in Portland and Seattle, clashes between protesters and police sometimes turned violent and were often followed by accusations of police brutality. During this hundred-day period, Portland police declared twenty-six riots.

Looking back at the three cities during the first hundred days after Floyd's death, we notice a split in the character of the protests. What started as one thing changed into something else. The first part of that time period seemed to be marked by protests that were more clearly focused on the messages of the Black Lives Matter movement. An early June BLM protest in Seattle featured a standoff between police and protesters with umbrellas, a tactic borrowed from demonstrators in Hong Kong.

100 DAYS OF PROTEST

Major events associated with the protests after the killing of George Floyd

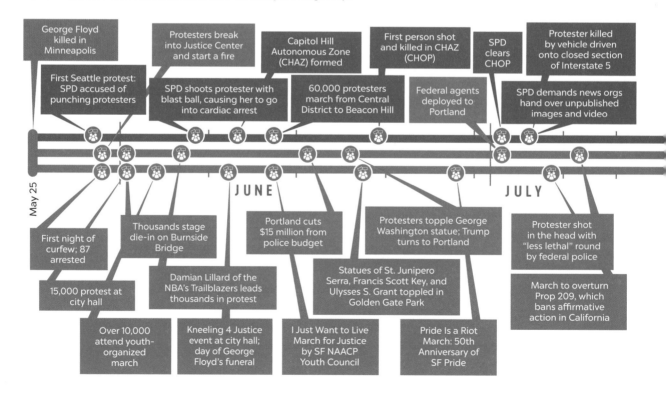

- George Floyd killed in Minneapolis
- First Seattle protest: SPD accused of punching protesters
- Protesters break into Justice Center and start a fire
- SPD shoots protester with blast ball, causing her to go into cardiac arrest
- Capitol Hill Autonomous Zone (CHAZ) formed
- 60,000 protesters march from Central District to Beacon Hill
- First person shot and killed in CHAZ (CHOP)
- Federal agents deployed to Portland
- SPD clears CHOP
- Protester killed by vehicle driven onto closed section of Interstate 5
- SPD demands news orgs hand over unpublished images and video

May 25 · JUNE · JULY

- First night of curfew; 87 arrested
- Thousands stage die-in on Burnside Bridge
- Portland cuts $15 million from police budget
- Protesters topple George Washington statue; Trump turns to Portland
- Protester shot in the head with "less lethal" round by federal police
- 15,000 protest at city hall
- Damian Lillard of the NBA's Trailblazers leads thousands in protest
- Statues of St. Junipero Serra, Francis Scott Key, and Ulysses S. Grant toppled in Golden Gate Park
- March to overturn Prop 209, which bans affirmative action in California
- Over 10,000 attend youth-organized march
- Kneeling 4 Justice event at city hall; day of George Floyd's funeral
- I Just Want to Live March for Justice by SF NAACP Youth Council
- Pride Is a Riot March: 50th Anniversary of SF Pride

July through early September, particularly in Portland and Seattle, marked an increasingly violent period where the focus of the protests drifted from BLM to more of a focus on defunding police departments. The tumult catalyzed the resignation of the police chiefs in both Portland and Seattle.

Over the summer, Portland and Seattle protesters became a specific focus of President Donald Trump's ire. On June 8, in Seattle, protesters declared an autonomous zone that became known as the Capitol Hill Autonomous Zone (CHAZ) or the Capitol Hill Occupied/Organized Protest (CHOP). In early July, federal agents were deployed to Portland, which escalated protests and violence there and amplified Portland's image as a protest city. Later that month, federal agents were also sent to Seattle and on July 1 police cleared the protester-organized CHAZ.

On September 21, three weeks after this time line ends, the Department of Justice officially identified Portland, Seattle, and New York City as anarchist jurisdictions. The labels came in support of President Trump's missive "Memorandum on Reviewing Funding to State and Local Government Recipients That Are Permitting Anarchy, Violence, and Destruction in American Cities."

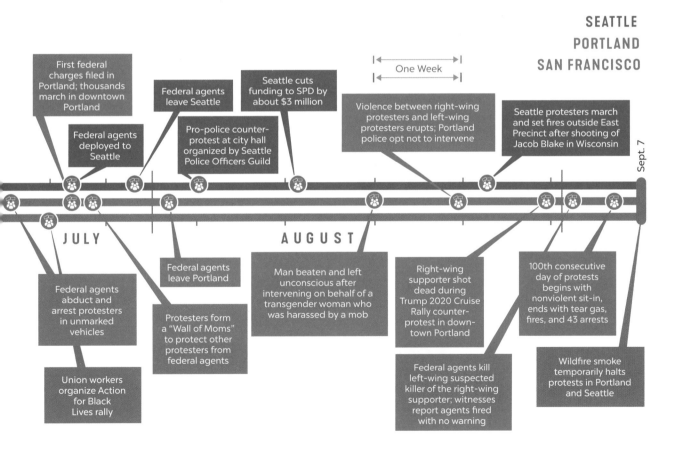

SEATTLE
PORTLAND
SAN FRANCISCO

One Week

First federal charges filed in Portland; thousands march in downtown Portland

Federal agents deployed to Seattle

Federal agents leave Seattle

Pro-police counter-protest at city hall organized by Seattle Police Officers Guild

Seattle cuts funding to SPD by about $3 million

Violence between right-wing protesters and left-wing protesters erupts; Portland police opt not to intervene

Seattle protesters march and set fires outside East Precinct after shooting of Jacob Blake in Wisconsin

Sept. 7

JULY

AUGUST

Federal agents abduct and arrest protesters in unmarked vehicles

Federal agents leave Portland

Protesters form a "Wall of Moms" to protect other protesters from federal agents

Man beaten and left unconscious after intervening on behalf of a transgender woman who was harassed by a mob

Right-wing supporter shot dead during Trump 2020 Cruise Rally counter-protest in downtown Portland

100th consecutive day of protests begins with nonviolent sit-in, ends with tear gas, fires, and 43 arrests

Union workers organize Action for Black Lives rally

Federal agents kill left-wing suspected killer of the right-wing supporter; witnesses report agents fired with no warning

Wildfire smoke temporarily halts protests in Portland and Seattle

San Francisco, Portland, and Seattle all have histories with high-profile protests. San Francisco's still well-known for protests that began in the 1960s opposing the Vietnam War and demonstrations for gay rights that made national headlines in the 1970s. On May 21, 1979, a riot occurred outside San Francisco's city hall as crowds reacted angrily to the lenient sentence given to the killer of Harvey Milk, who was the first openly gay elected official in the history of California.

In 1999, Seattle hosted a conference of the World Trade Organization which attracted tens of thousands of protesters. Violent confrontations and the tear-gassing of many protesters over a four-day period is sometimes called the Battle of Seattle and is still noted for bringing the anti-globalization movement to the attention of US public.

Portland protesters staged such vociferous demonstrations against the Gulf War in 1991 that a staffer of President of George H. W. Bush dubbed the city Little Beirut. In addition to maligning the capital of Lebanon, this moniker presaged the portrayal of Portland as an anarchic city of rioters by the Trump administration in 2020.

The maps on these pages depict some of the frequent and notable locations of protests in San Francisco, Portland, and Seattle between May 25 and September 7, 2020, the span of the time line on the previous pages.

SEATTLE

BALLARD BRIDGE
Marchers closed bridge
August 26 for 8 minutes
and 46 seconds

MAGNUSON PARK
Starting point for June 6
march to University Village

WEST PRECINCT
Site of many protests;
city surrounds building
in cement barricades

INTERSTATE 5
Protesters closed a
downtown stretch for 19
straight days in June/July

WESTLAKE CENTER
Meeting place for marches
and demonstrations

CHAZ/CHOP
Focal point for protests
until cleared on July 1

CITY HALL
Location of
several protests

JUDKINS PARK
Site of The Next Steps
rally on Juneteenth

BEACON HILL
Destination for March
of Silence on June 12

**SEATTLE POLICE
OFFICERS GUILD**
August 16 protest
declared a riot, 18
arrested

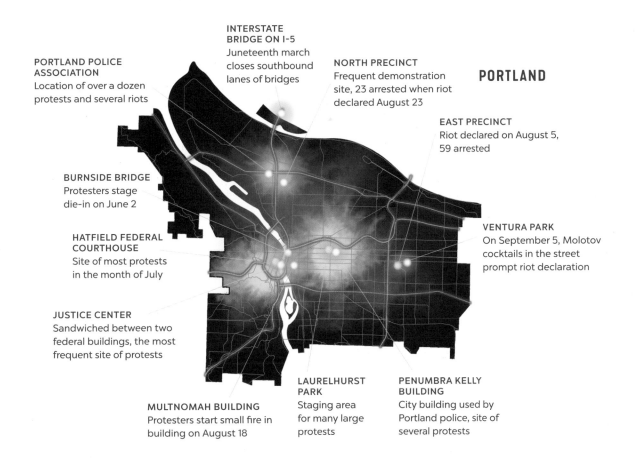

PORTLAND POLICE ASSOCIATION
Location of over a dozen protests and several riots

INTERSTATE BRIDGE ON I-5
Juneteenth march closes southbound lanes of bridges

NORTH PRECINCT
Frequent demonstration site, 23 arrested when riot declared August 23

PORTLAND

EAST PRECINCT
Riot declared on August 5, 59 arrested

BURNSIDE BRIDGE
Protesters stage die-in on June 2

HATFIELD FEDERAL COURTHOUSE
Site of most protests in the month of July

VENTURA PARK
On September 5, Molotov cocktails in the street prompt riot declaration

JUSTICE CENTER
Sandwiched between two federal buildings, the most frequent site of protests

MULTNOMAH BUILDING
Protesters start small fire in building on August 18

LAURELHURST PARK
Staging area for many large protests

PENUMBRA KELLY BUILDING
City building used by Portland police, site of several protests

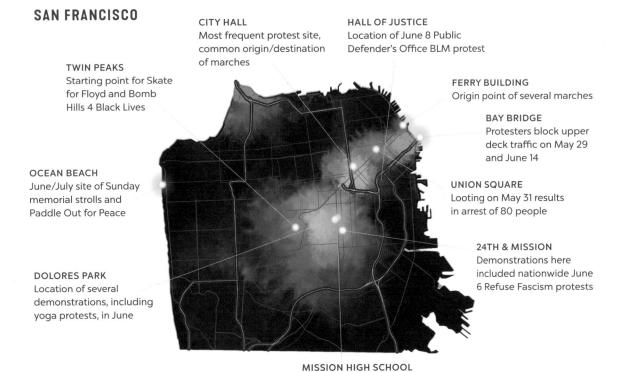

SAN FRANCISCO

CITY HALL
Most frequent protest site, common origin/destination of marches

HALL OF JUSTICE
Location of June 8 Public Defender's Office BLM protest

TWIN PEAKS
Starting point for Skate for Floyd and Bomb Hills 4 Black Lives

FERRY BUILDING
Origin point of several marches

BAY BRIDGE
Protesters block upper deck traffic on May 29 and June 14

OCEAN BEACH
June/July site of Sunday memorial strolls and Paddle Out for Peace

UNION SQUARE
Looting on May 31 results in arrest of 80 people

24TH & MISSION
Demonstrations here included nationwide June 6 Refuse Fascism protests

DOLORES PARK
Location of several demonstrations, including yoga protests, in June

MISSION HIGH SCHOOL
Origin point of several marches to city hall

UNKNOWN MEASURES

THE MAPS

Maps are representations. With maps, we can create physical representations of the world—mountains will be made to look like mountains. Maps can also create conceptual representations of the world. In this case, areas of high population might be made to look like mountains. These impressionistic maps juxtapose the elevations and populations of our three cities. Across the three cities, the highest points tend to have low populations.

San Francisco has a reputation for being hilly, but hillier still are the Marin Headlands to the north. A look at the population map shows that the headlands, and other high points in the area, have smaller populations. San Francisco's population map does a better job of revealing the city. The map shows peaks north of downtown in the northeastern parts of the city and a valley to the west, which is Golden Gate Park.

Portland turns out to be pretty flat, aside from the West Hills and a few volcanic remnants. As might be expected from a city located in a river valley, elevation is low. The Pearl District, the spike to the west of downtown, has the densest population in the city. A look at both maps reveals that, generally, there is an inverse relationship between elevation and population.

Seattle's elevation map shows a dramatic interplay between land and water, with sharp hills abutting the water. Imagine how much more this map would be exaggerated had Seattle's many regrades not taken place. The population map shows a spike of high population that corresponds with some of the flatter areas of the city. The high elevations of the west part of the city are largely single-family homes and host relatively little population.

THE INSPIRATION

These maps were inspired by the radio pulsar data graph on the cover of Joy Division's groundbreaking album *Unknown Pleasures*. Band members found it in *The Cambridge Encyclopaedia of Astronomy*, excerpted from the 1970 doctoral dissertation of radio astronomer Harold Craft. The plot examines the frequency of the first pulsar discovered. Designer Peter Saville inverted the original black lines and white background to make the album's famous cover.

The image, tightly associated with the band and album, has become an icon in its own right. The cover art has been used for T-shirts, hoodies, tattoos, and memes, and reaches a global audience. Adidas, Carhartt, Dr. Martens, New Balance, Nike, and Supreme have all produced merchandise with its image.

What explains the longevity of these jagged lines in an era that is so image-saturated? Certainly, the impact of Joy Division continues to play its part. *Unknown Pleasures*, the only album released during the lifetime of vocalist and lyricist Ian Curtis, is key to the band's legacy.

The graph alone is compelling, more so when associated with Joy Division's gritty atmospheric sound. There is something sparse and somber about this image.

The *Unknown Pleasures* graph is sometimes mistaken for a map. Although the original is not a map, our versions are. What began in an academic text became album art. With our update, the image is once again in a text. Nearly fifty years after Craft first published it, the graphic still helps to tell new stories.

SAN FRANCISCO

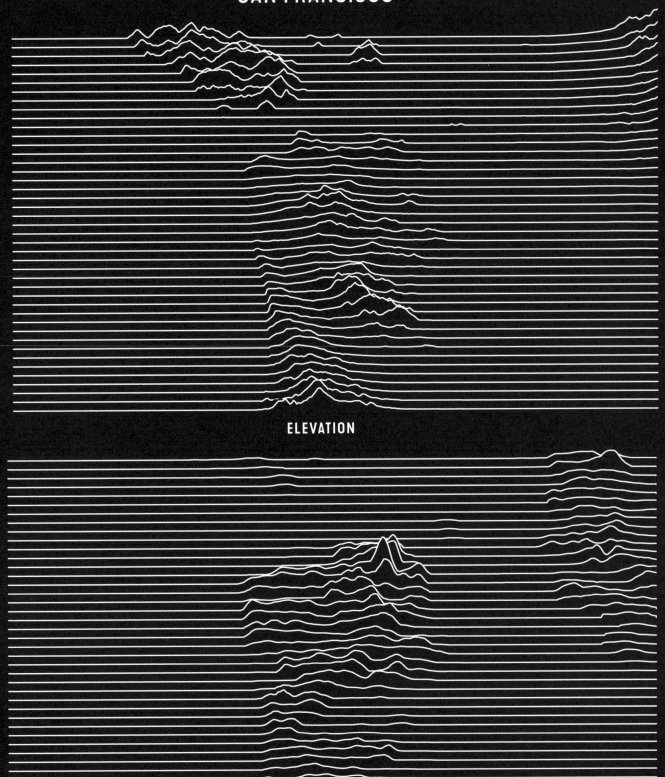

ELEVATION

POPULATION

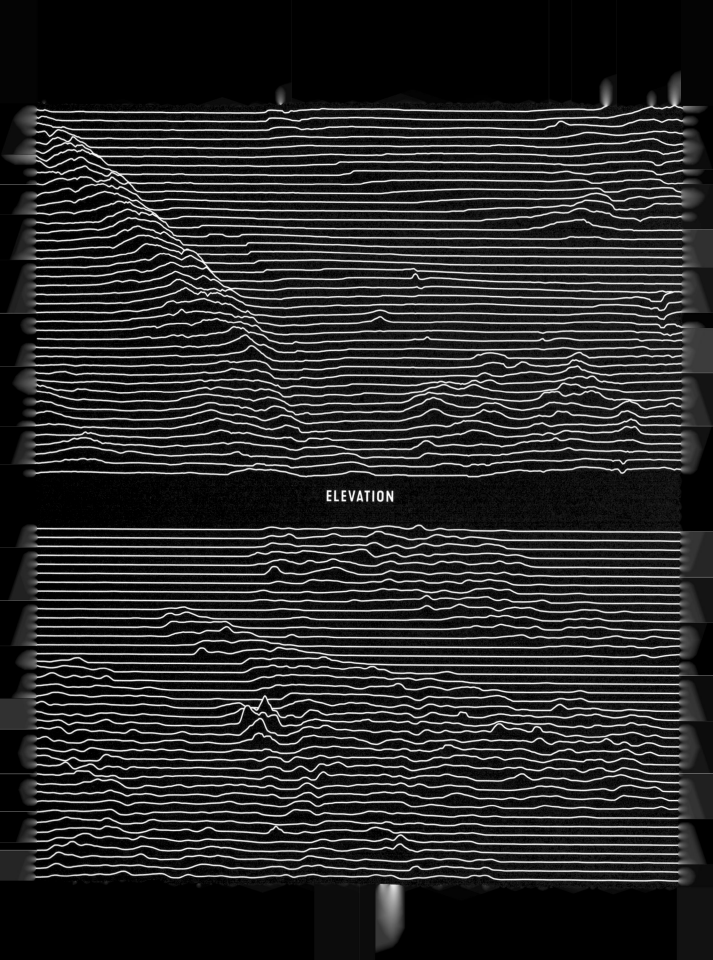

ELEVATION

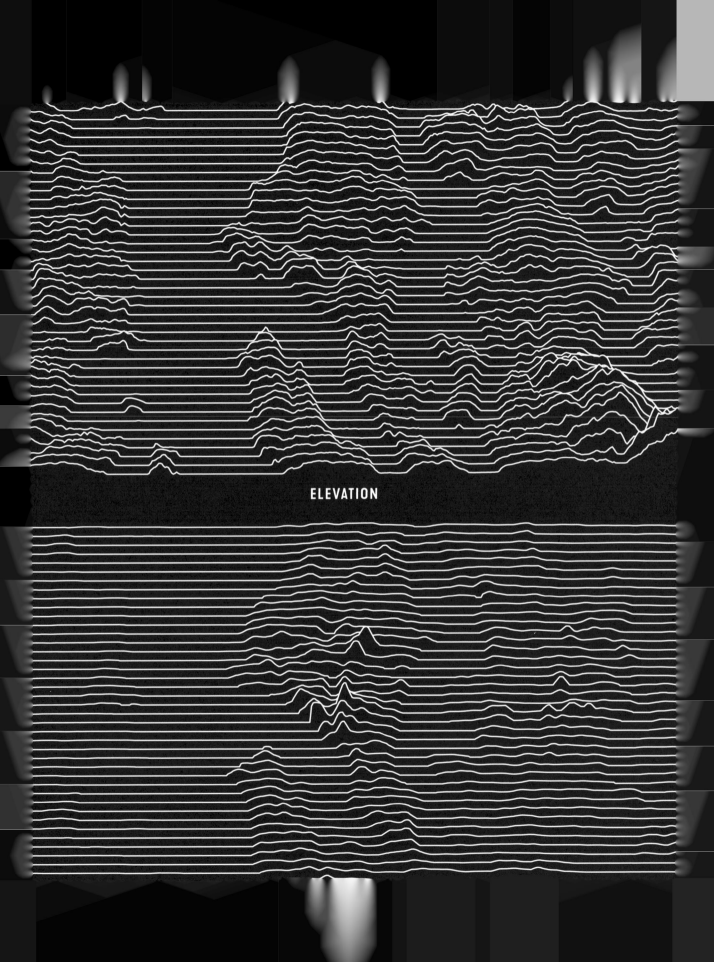

ELEVATION

MORE THAN WORDS: LANGUAGES

San Francisco, Portland, and Seattle each have their own geographies of spoken languages. The maps on these pages examine the top five most spoken languages (other than English) in each metro area (listed in order of most spoken, reading left to right and top to bottom) as of 2016. The data comes from the American Community Survey, an ongoing demographic data-collection project of the United States Census.

Nationally, the most spoken language besides English is Spanish, with nearly 40 million speakers. Chinese is the second in the United States, with 3.2 million speakers. In order of number of speakers, the list continues with Tagalog, Vietnamese, French (including Creole), Arabic, Korean, German, Russian, and Haitian.

Unlike many countries in the world, the United States has no official language or languages. Some states, however, have adopted English as their official language. Such adoptions have sometimes meant replacing bilingual education programs with English-only immersion programs, which happened in California from 1998 to 2016.

Spanish, Chinese, and Vietnamese all placed in the top five most spoken languages in San Francisco, Portland, and Seattle. Spanish ranked first in both Portland and Seattle. Chinese placed first as the most spoken language in San Francisco, while Spanish placed second. As of 2016, San Francisco had approximately 143,000 Chinese speakers and 89,000 Spanish speakers.

In San Francisco, Chinese and Spanish speakers are spread widely throughout the city. Comparatively, there is a higher density of Chinese speakers than Spanish speakers in the Richmond and Sunset Districts on the east side of the city. There is a higher density of Spanish speakers in the Mission. Tagalog speakers are more concentrated in the neighborhoods in the south of the city, including the Excelsior and the Outer Mission, with another concentration north of downtown. Russian speakers are most densely concentrated in the Richmond District, with other concentrations in the north and southwest of the city. There are pockets of Vietnamese speakers north of downtown and in the southeast of the city, such as the Portola neighborhood.

In Portland, there are high numbers of Spanish speakers in the north and east parts of the city, as well as the east and west sides of the metro area. Vietnamese, Chinese, and Russian speakers are highly concentrated in East Portland. There is also a notable concentration of Chinese speakers west of Portland around Bethany. Russian speakers are most concentrated on the east side of the Portland metro area, while Vietnamese speakers are more concentrated in Hillsboro to the west. North Portland has the highest concentration of speakers of African languages, with some concentrations in northeast and southeast as well.

Seattle's Spanish speakers are spread throughout the city and metro, with concentrations in North Seattle, south of Seattle, and in a stretch from Capitol Hill to Beacon Hill through Rainier Valley. The metro area also has a widely diffused pattern of speakers with a concentration south of Seattle. High densities of Chinese speakers are found in the University District and a band from the International District through Rainier Valley. At the Seattle metro level, the densest concentrations of Chinese speakers are found west of the city. Concentrations of Vietnamese, African language, and Tagalog speakers are also found in the southeastern part of the city. African language speakers are also concentrated south of the city near the airport.

SEATTLE

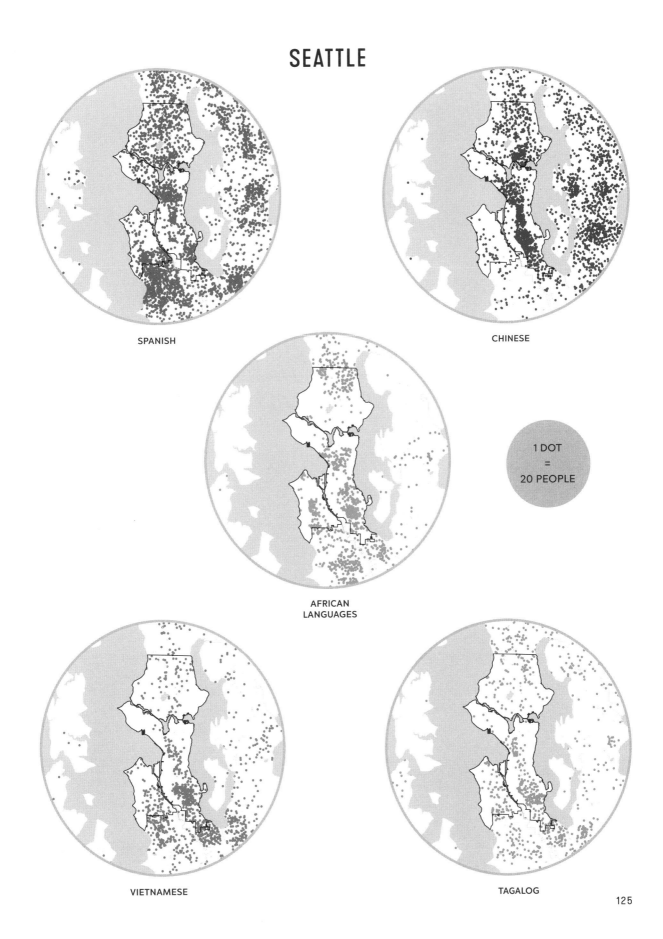

SPANISH

CHINESE

AFRICAN
LANGUAGES

1 DOT
=
20 PEOPLE

VIETNAMESE

TAGALOG

125

PORTLAND

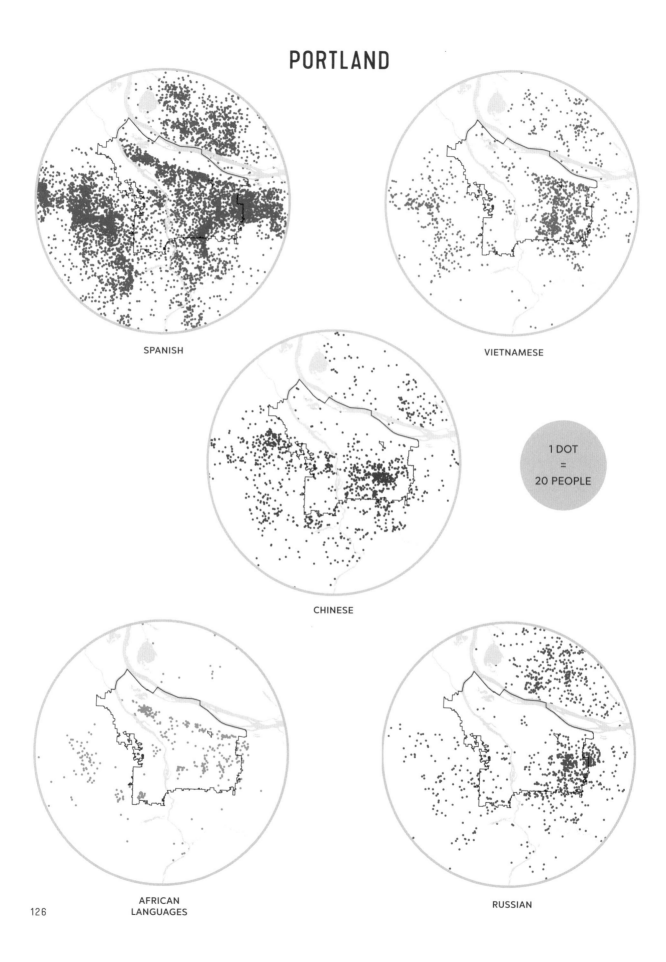

SPANISH

VIETNAMESE

1 DOT
=
20 PEOPLE

CHINESE

AFRICAN
LANGUAGES

RUSSIAN

SAN FRANCISCO

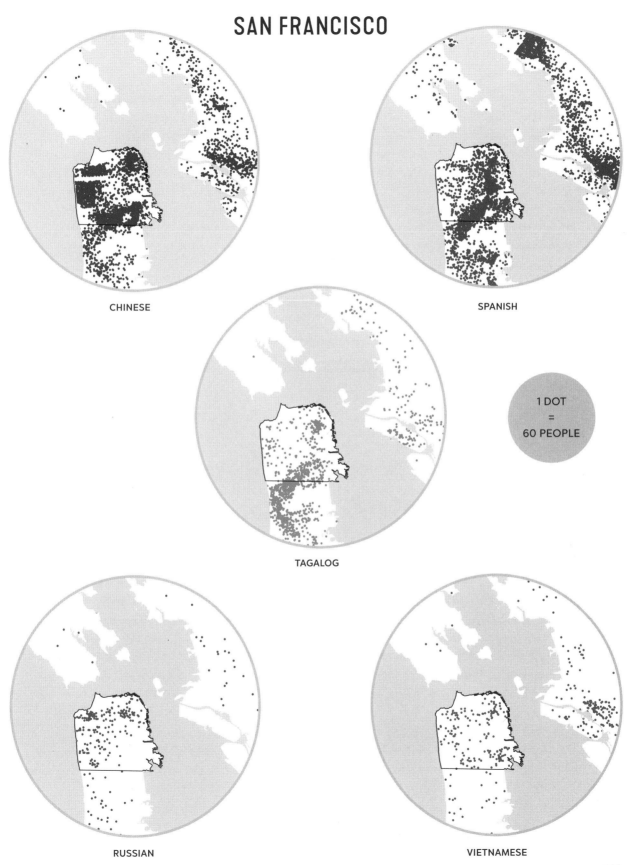

CHINESE

SPANISH

TAGALOG

1 DOT
=
60 PEOPLE

RUSSIAN

VIETNAMESE

CHINATOWNS AND JAPANTOWNS

Chinatowns and Japantowns (Nihonmachi) existed because of and in spite of discrimination leveled at immigrants from East Asia. Early Chinese and Japanese immigrants formed a community in Upper Left cities that served as a comfort for laborers and a welcome familiarity for subsequent waves of immigrants. By providing a place for rest, laundry, provisions, and worship, these districts generated a sense of belonging among Chinese and Japanese immigrants experiencing social and political intolerance. The stories of how each district offered fortitude, community, and home can be traced through the shifts of their boundaries.

CHINATOWN

In 1849, Chinese men began arriving in San Francisco lured by the gold rush. Often motivated by poverty, first-wave immigrants set sail for America with hopes of providing for their families. Far fewer women were allowed to migrate to the United States, and many who did were forced into prostitution. White laborers viewed this "bachelor society" of Chinese laborers as a threat to their jobs. This xenophobia resulted in the 1882 Chinese Exclusion Act, which prohibited immigration for Chinese laborers. The exclusion of Chinese workers would not be officially lifted until 1943.

San Francisco's early Chinese entrepreneurs settled near the city center to create a business district. This small village grew to twelve blocks of retail stores, joss houses, lodging, herbalists, restaurants, and schools.

San Francisco's 1906 earthquake and subsequent fire destroyed much of downtown, including Chinatown. Chinese leaders influenced the city's decision to rebuild in place rather than relocate. The opportunity to rebuild was used to redesign to a

tourist-friendly aesthetic of pagodas and ornamentation, which one can still see today.

Seattle's early Chinese immigrants settled in the area now known as Pioneer Square. They remained there until 1886, when a mob blamed the three hundred Chinese residents for labor shortages. Rioters pulled residents from their homes and forced most to board a ship set for San Francisco. After the Great Seattle Fire of 1889, the few Chinese who remained in Seattle rebuilt a second Chinatown on blocks along both sides of South Washington Street. Later, the Jackson Regrade project would displace businesses and by 1910, the Chinese community had moved to King Street, establishing Seattle's current Chinatown location.

Seattle's Chinatown was incorporated into the International District in 1951, to be inclusive of Chinese, Japanese, Filipino, and other communities represented in the district's broad ethnic demographic.

Portland's two early Chinatowns were split between urban and rural. In the 1890s, rural Chinatown, known as the Chinese Vegetable Gardens, was located where Providence Park is today.

The urban Chinatown of Portland's early history began downtown in the late 1880s. Chinese residents and businesses then moved north of Burnside due to continuous flooding of the Willamette River. A seven-block northern Chinese business district began to form and became known as New Chinatown. A flood in 1894 displaced most Old Chinatown establishments. Conversions of parcels to white ownership and development further pushed Chinese owners to establish themselves in New Chinatown. Portland's New Chinatown was later designated as the New Chinatown/Japantown Historic District, representing the interrelated nature of these two historic enclaves.

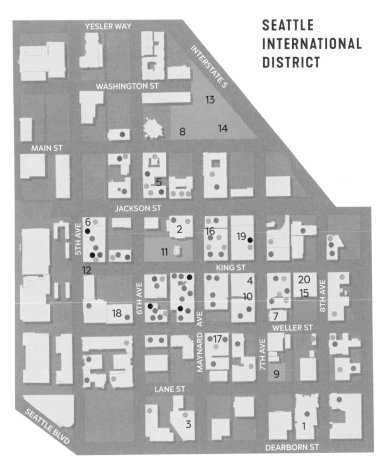

SEATTLE INTERNATIONAL DISTRICT

1. Chinatown Community Center
2. Chinatown Int'l Dist. Business Improvement Area
3. Chinese Community Bulletin Board
4. Chinese Information and Services Center
5. Chiyo's Garden
6. Choeizan Enkyoji Nichiren Buddhist Temple
7. Chong Wa Benevolent Association
8. Danny Woo Community Garden
9. Donnie Chin International Children's Park
10. Gee How Oak Tin Family Association
11. Hing Hay Park
12. Historic Chinatown Gate
13. Kobe Park
14. Kobe Terrace
15. Korean American Historical Society
16. NW Asian Weekly
17. Seattle Chinese Services
18. Soy Source
19. Taiwan Buddhist Tzuchi
20. Wing Luke Museum

- Japanese restaurants
- Chinese restaurants
- Other Asian restaurants
- Retail businesses
- Bars
- Martial arts

JAPANTOWN

The decrease of Chinese immigrants through the 1882 Chinese Exclusion Act led to a push for Japanese labor. The years following the exclusion saw a massive uptick in Japanese populations in each of the Upper Left cities. However, many Japanese immigrants found themselves limited to the same racially restrictive covenants and employment discrimination experienced by Chinese immigrants, causing Japanese enclaves to establish close to Chinatowns.

San Francisco's first Japanese immigrants, or *issei*, moved into Chinatown, as well as in the working-class district, South of Market. Issei, largely men who fled Japan's depressed economy, began arriving as early as 1869.

As in Chinatown, the 1906 earthquake flattened Japanese settlements. The community was re-established in the South Park and Western Addition areas. The small Nihonmachi (or Japanese Town) in South Park catered to Japanese travelers passing through the port until the 1930s. The Western Addition/Fillmore District emerged as a second Nihonmachi, one that exists to this day.

Issei began to arrive in Seattle in the 1890s, establishing a fifteen-block Nihonmachi to the north and east of Chinatown. Although many Japanese farmers would settle in rural areas, traveling into Pike Place Market to supply more than half of Seattle's fruits, vegetables, and dairy until the 1920s. Discrimination of land ownership curbed Japanese-owned agriculture, while businesses in the city fared better. The growth of families and

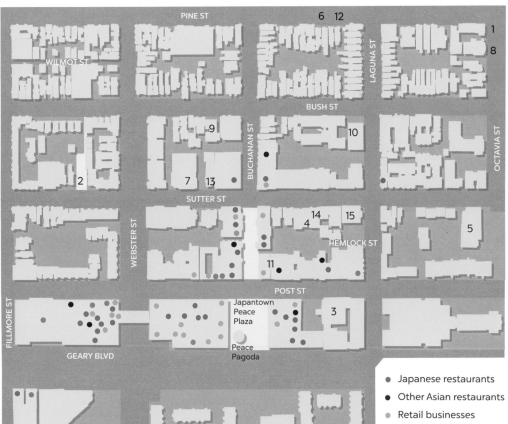

1. Buddhist Church of San Francisco
2. Cottage Row Park
3. Hotel Kabuki
4. Japanese American Citizens League
5. Japanese American National Library
6. Japanese Community Youth Council
7. Japanese Cultural and Community Center
8. Japanese Language School
9. Kinmon Gakuen Golden Gate Institute
10. Konko Church
11. National Japanese American Historical Society
12. Nichiren Hokke Buddhist Church
13. Nihonmachi Little Friends
14. Northern California Cherry Blossom Festival
15. Soto Zen Mission of San Francisco—Sokoji

- Japanese restaurants
- Other Asian restaurants
- Retail businesses
- Entertainment

relative economic stability helped this settlement become the second-largest Japanese community on the West Coast (after San Francisco).

Portland's early Nihonmachi formed between Broadway and the Willamette River, bounded by Burnside, although a second core neighborhood is thought to have emerged in the southwest quarter, stretching up to Southwest Montgomery. Similar to Seattle, Portland's issei opened hotels, laundries, bathhouses, barbershops, and gambling houses to cater to laborers and incoming immigrants. Not subject to Chinese immigration laws, Japanese women could immigrate as the wives of laborers. Their presence encouraged family-oriented lifestyles and resulted in schools, beauty shops, and candy shops that catered to women and children.

This early Nihonmachi has been remembered as a safe and comfortable place.

In 1942, during WWII, Executive Order 9066 was adopted, forcing Japanese Americans and Japanese immigrants into internment camps. The majority of Seattle and Portland's Japanese were incarcerated in Minidoka, Idaho. Many of San Francisco's Japanese residents were relocated to Topaz War Relocation Center in Utah, through 1945.

During internment, San Francisco's Nihonmachi became a cultural and economic space for many African Americans. San Francisco's Japanese community returned to their neighborhood after release from internment. During the urban renewal period of the 1960s, a path was cleared for the development of the Japan Center, which opened in 1968. Unlike Nihonmachi in Seattle and Portland, San Francisco's Nihonmachi still endures.

Many abandoned homes and businesses in Seattle's Nihonmachi were soon reoccupied by African Americans on military duty or filling wartime jobs. Upon returning from internment, Japanese families were displaced from the original Nihonmachi and were unable to reformulate a distinctly Japanese neighborhood. Instead, the community gathered through social and cultural events. Significant buildings such as the Panama Hotel and the Uwajimaya Asian market are landmarks within the historic Japantown district, while the Japanese Cultural and Community Center on the edge of the International District is a hub to support language and cultural traditions for the Japanese community. Today, Nihonmachi is integrated into the International District.

During the internment period, Portland's Nihonmachi was largely repopulated by Chinese business owners and residents. While some hoteliers and other businesspeople re-established their businesses upon release from incarceration, Nihonmachi never returned to the richly Japanese district it was before internment. Today, Japanese cultural life is kept alive through events hosted by social groups and religious establishments, as well as by the Oregon Nikkei Legacy Center in the New Chinatown/Japantown Historic District.

THE NEW NEW

In recent years, Chinese and Japanese communities have branched out from historic enclaves, in some cases forming new concentrations. San Francisco has one of the highest Chinese populations in the country, with concentrations in Chinatown, the Richmond, the Sunset, and beyond. San Francisco's Japan Center continues to be a node for the Japanese community. Many of Seattle's Chinese community call North Beacon Hill home, but many others have settled across Seattle into the surrounding metro areas. Portland's Chinatown has almost disappeared entirely. East Portland near Eighty-Second Avenue currently serves as home to many residents of East and Southeastern Asian heritage.

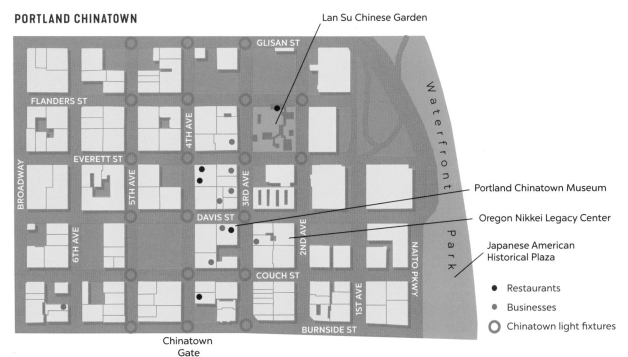

PORTLAND CHINATOWN

Lan Su Chinese Garden

GLISAN ST

FLANDERS ST

4TH AVE

EVERETT ST

BROADWAY

5TH AVE

3RD AVE

DAVIS ST

6TH AVE

2ND AVE

COUCH ST

1ST AVE

NAITO PKWY

BURNSIDE ST

Chinatown
Gate

Waterfront Park

Portland Chinatown Museum

Oregon Nikkei Legacy Center

Japanese American
Historical Plaza

● Restaurants
● Businesses
○ Chinatown light fixtures

FOR WHERE THE BELL TOLLS

Although we might not think about it this way, part of the landscape of a city is made up of sound, or soundscapes. Some common elements of the soundscape are harsh, such as piercing ambulance sirens, music blasting from passing cars, and loud barking dogs. Other sounds are softer, like light rain falling on the sidewalk, the rustle of leaves in a breeze, and water running from a fountain. Some sounds come on a predetermined schedule. Train horns, which can be heard from miles away, boom predictably, day and night.

Another prevalent and predictable sound heard throughout cities is the peal of church bells. While not as loud as a train whistle, church bells can be heard from blocks away or further. Although there are synagogues and mosques in Upper Left cities, they are less represented in each city's soundscape, as compared to Catholic, Episcopalian, and Presbyterian churches. Regardless of religious affiliation, the chime of bells from Christian churches will likely be a regular feature of the city-dweller's soundscape.

In modern times, the primary use for church bells is to signal the commencement of church activities. Additionally, church bells can be used in celebration of special dates or events such as weddings and holidays. Bells are also used in remembrance of lost community members.

On August 25, 2019, tolling of church bells became part of a powerful display of national remembrance and healing. The National Park Service invited faith communities from around the country to participate in commemorating the four hundredth anniversary of enslaved Africans who arrived in Virginia on August 25, 1619. Church bells rang for one minute at three in the afternoon in a unique act of remembrance and healing.

On these pages we represent the soundscape of church bells in each city. There is no central database anywhere (that we could find) with church bell data. We collected this data ourselves. In conducting this research, we found that while some churches still have active bells, it is more common for a church to broadcast a recording of bells. Still, human bell ringers are required in a few locations.

There are many reasons why churches choose not to use or maintain church bells. For some, the cost of upkeep is prohibitive. Other churches removed church bells for earthquake-safety reasons.

While some people take comfort in the sound of church bells, others find the noise so disruptive that they initiate noise-nuisance litigation. Such was the case with Saints Peter and Paul Church, located in San Francisco's North Beach, which no longer uses actual bells. Instead, a recorded set of bells is broadcast through speakers. This broadcast was the basis of a noise complaint filed by an agitated neighbor living half a block from the church, who eventually sued the church in small-claims court for violation of San Francisco's noise ordinance. The ruling dismissed the complaint.

In some cases, not having a bell can be as interesting as having one. Saint Mark's Episcopal Cathedral in Seattle had original plans for a magnificent Gothic-style structure complete with rose windows and large carillon towers. Because of the stock market crash of 1929 and the Great Depression, portions were left incomplete, the towers never built. The unfinished nature of the structure has become central to its identity.

SAN FRANCISCO

The bells at Holy Trinity Cathedral were removed before the 1906 fire that destroyed the Cathedral. The original bells were later installed in the rebuilt tower.

Grace Cathedral has a forty-four-bell carillon that can be played either live from a keyboard or programmed automatically.

One day in 2016 at Saint Aidan's Episcopal Church, a baseball somehow became lodged between the bottom of the belfry structure and the bell itself, silencing the bell ever since.

The bell at the Episcopal Church of Saint John the Evangelist is named Lady Julian after both the street on which they are located and Julian of Norwich, an English anchorite of the Middle Ages.

- ◎ Church with bell that rings
- ○ Church with bell not in use
- · Church with no bells
- · Not sure

PORTLAND

First Presbyterian Church has Portland's oldest bell, first installed in 1863 and moved to the current church in 1890.

The Catholic church Saint Elizabeth of Hungary has a large bell that was moved from South Portland's Saint Lawrence Catholic Church, which was demolished in the 1960s to make way for the Marquam Bridge (I-5).

The bell tower of the First Congregational United Church of Christ was Portland's tallest building for more than sixty years, with construction completed in 1895; the bell has rung ever since.

Portland State University, which has no bell tower, plays a lovely bell recording from a loudspeaker atop the student union building.

◎ Church with bell that rings

○ Church with bell not in use

● Church with no bells

● Not sure

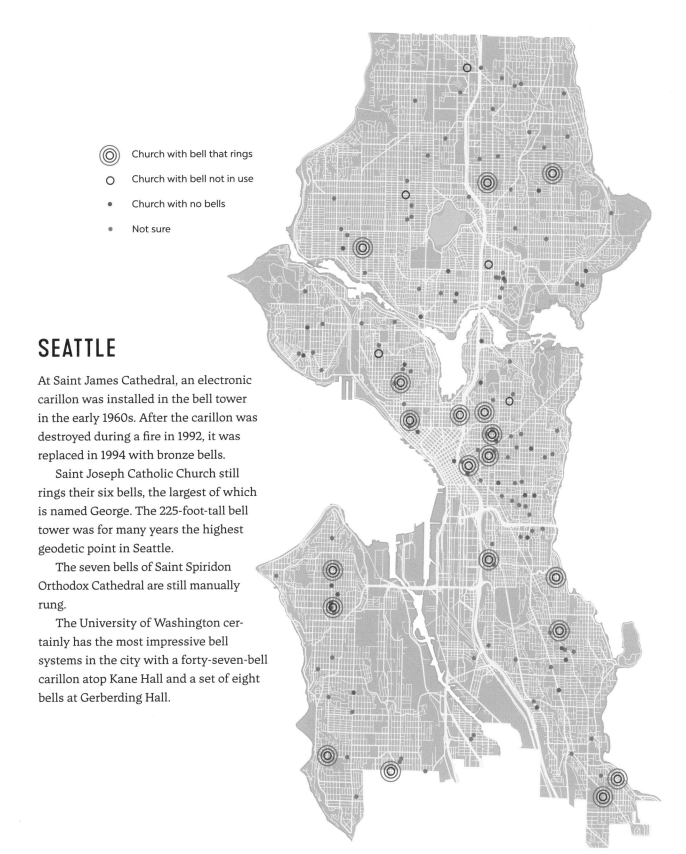

SEATTLE

At Saint James Cathedral, an electronic carillon was installed in the bell tower in the early 1960s. After the carillon was destroyed during a fire in 1992, it was replaced in 1994 with bronze bells.

Saint Joseph Catholic Church still rings their six bells, the largest of which is named George. The 225-foot-tall bell tower was for many years the highest geodetic point in Seattle.

The seven bells of Saint Spiridon Orthodox Cathedral are still manually rung.

The University of Washington certainly has the most impressive bell systems in the city with a forty-seven-bell carillon atop Kane Hall and a set of eight bells at Gerberding Hall.

Church with bell that rings

Church with bell not in use

Church with no bells

Not sure

IV. COMMERCE

CITIES ARE BIG BUSINESS. San Francisco, Portland, and Seattle were established, first and foremost, as ports. All three cities boomed in conjunction with gold rushes and have ridden the boom and bust roller coaster of capitalism ever since.

Businesses come and go, as do entire industries. The cycle of obsolescence is shorter and shorter. The most essential piece of business equipment for years was the typewriter. Now we buy and work on computers that, if you wish to keep up to speed, must be replaced every few years. Over the past decades, shipping industries, which were critical to local economies from their inception, have declined in the ports of the Upper Left. During that same time, a new domestic product has become a commercial phenomenon in each city: cannabis.

In business, there are costs and benefits. For the benefit of faster transportation, cities have borne the cost of environmental pollution. Ever evolving, transportation networks connect cities nationally and internationally for business and pleasure. For example, an international airport (a hub) is critical for a city's global standing.

In this chapter, we delve into the economic dimensions of San Francisco, Portland, and Seattle.

TRADE-OFFS

San Francisco, Portland, and Seattle all owe their existence to their ports. During World War II, each city became the location of large shipbuilding industries. After the war, trade expanded rapidly. Even today, each city still has a functioning port. For value of trade, Seattle's seaport ranks thirteenth in the country, compared to thirty-second place for Portland and forty-seventh for San Francisco. In 2018, total seaport trade was valued at about $5.6 billion for San Francisco, $12.4 billion for Portland, and $30.2 billion for Seattle. A lot of money, but not quite the $299 billion of trade that went through the Port of Los Angeles the same year.

Today, imports and exports not only flow through seaports but through airports as well. An authority independent of the Port of San Francisco controls the San Francisco International Airport (SFO). This is different from Portland and Seattle, where the authorities that operate the seaports also operate the airports. In 2018, the value of all trade passing through SFO was $66.4 billion. It was $1.98 billion for the Portland International Airport (PDX) and $30.9 billion for the Seattle-Tacoma International Airport (Sea-Tac).

San Francisco, a city on the ocean, has hosted very little shipping traffic since the 1960s and the advent of container ships. The fortune of ports is now very tied up with global container shipping. Shipping largely shifted to the Port of Oakland. By value, the overwhelming majority of trade now goes through SFO.

Top exports for SFO are semiconductor parts and machinery, computer chips, cell phones, and computers. The list of top imports to SFO is quite similar—computer chips, computer parts, cell phones, and computers.

Portland is located on both the Columbia and Willamette Rivers, about eighty miles from the Pacific Ocean. Automobiles are the number-one

IMPORTS BY VALUE FOR SFO INTERNATIONAL AIRPORT

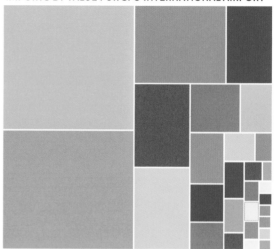

Top ten: 1. Computer chips; 2. Computer parts; 3. Cell phones; 4. Computers; 5. Machinery & parts for semiconductor manufacturing; 6. Value added to a returned import; 7. Unrecorded media for audio; 8. Plasma, vaccines, blood; 9. Medical instruments; 10. Photosensitive semiconductors

IMPORTS BY VALUE FOR PORT OF SAN FRANCISCO

Top ten: 1. Electric storage batteries; 2. Oil; 3. Motor vehicles; 4. Coffee; 5. Motor vehicle parts; 6. Computers; 7. Gasoline & other fuels; 8. Machinery & parts for semiconductor manufacturing; 9. Furniture; 10. Power supplies & transformers

*Note: graphics do not show total import value

IMPORTS BY VALUE FOR PDX INTERNATIONAL AIRPORT

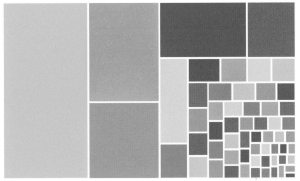

Top ten: 1. Misc. chemicals used for disc wafers; 2. Value added to a returned import; 3. Machinery & parts for semiconductor manufacturing; 4. Hydrogen, raw gases; 5. Medical instruments; 6. Medical technology; 7. Knives; 8. Computer chips; 9. Misc. machine parts; 10. Cell phones

IMPORTS BY VALUE FOR PORT OF PORTLAND

Top ten: 1. Motor vehicles; 2. Gasoline & other fuels; 3. Computers; 4. Machinery & parts for semiconductor manufacturing; 5. Nitrogenous fertilizers; 6. Value added to a returned import; 7. Medical technology; 8. Portland cement; 9. Misc. chemicals used for disc wafers; 10. Computer chips

IMPORTS BY VALUE FOR SEA-TAC INTERNATIONAL AIRPORT

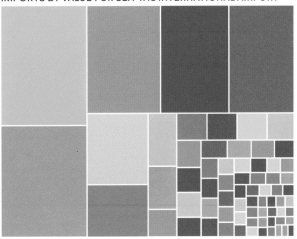

Top ten: 1. Aircraft engines & engine parts; 2. Machinery & parts for semiconductor manufacturing; 3. Value added to a returned import; 4. Unrecorded media for audio; 5. Computer chips; 6. Computers; 7. Defense-related aircraft & parts; 8. Cell phones; 9. Seats; 10. Misc. engine motor parts

IMPORTS BY VALUE FOR PORT OF SEATTLE

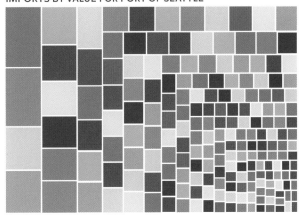

Top ten: 1. Furniture; 2. Defense-related aircraft & parts; 3. Motor vehicle parts; 4. Toys & games; 5. Seats; 6. Cell phones; 7. Athletic & other textile shoes; 8. Coffee; 9. TVs & computer monitors; 10. Plastic shoes

commodity both coming in and out of the Port of Portland. The city is the top exporter of vehicles on the West Coast.

More wheat is exported from Portland than from any other port in the country. This is due to Portland's Columbia River access to the Palouse, one of the biggest wheat belts in the country. The wheat comes from as far away as the Idaho border. The Port of Portland also controls the airports in nearby Hillsboro and Troutdale.

Seattle's location on Elliott Bay gives it access to the Puget Sound and, about 120 miles away, the Pacific Ocean. Unlike Portland and San Francisco, Seattle has no bridges to pass under between the city and the ocean, so there is no height limit for ships docking at the Port of Seattle.

The port of Seattle's top imports were furniture, defense-related aircraft parts, motor vehicles parts, and toys/games. The port's top exports were civilian aircraft parts, corn, soybeans, frozen fish, and potatoes. The Ports of Seattle and Tacoma have been in cahoots since 2014 when they formed the Northwest Seaport Alliance.

PORT OF PORTLAND

**TOP EXPORTS 2018
(5.1B TOTAL)**

Motor vehicles 33%

Wheat 22%

Potassic fertilizers 18%

Carbonates 15%

Commercial vehicles 2.9%

Soybeans 2.3%

Ethyl alcohol 2.1%

Oil 1.7%

Scrap iron and steel 1.4%

Tractors 0.74%

PORT OF SEATTLE

**TOP EXPORTS 2018
(9.1B TOTAL)**

Civilian aircraft and parts 12.2%

Corn 6.2%

Soybeans 5.9%

Frozen fish 4.5%

Potatoes 4.4%

Gasoline and other fuels 3.7%

Hay 3.1%

Fresh apples and pears 2.2%

Misc. articles of stone 1.9%

Beans and peas 1.7%

PORT OF SAN FRANCISCO

**TOP EXPORTS 2018
(1.7B TOTAL)**

Motor vehicles 71%

Gasoline and other
fuels 20%

Petroleum products 5.5%

Rice 1.3%

Nuts 0.95%

Defense-related aircraft and
parts 0.2%

Medical instruments 0.18%

Oils derived from coal 0.17%

Jewelry 0.16%

Seeds for sowing 0.13%

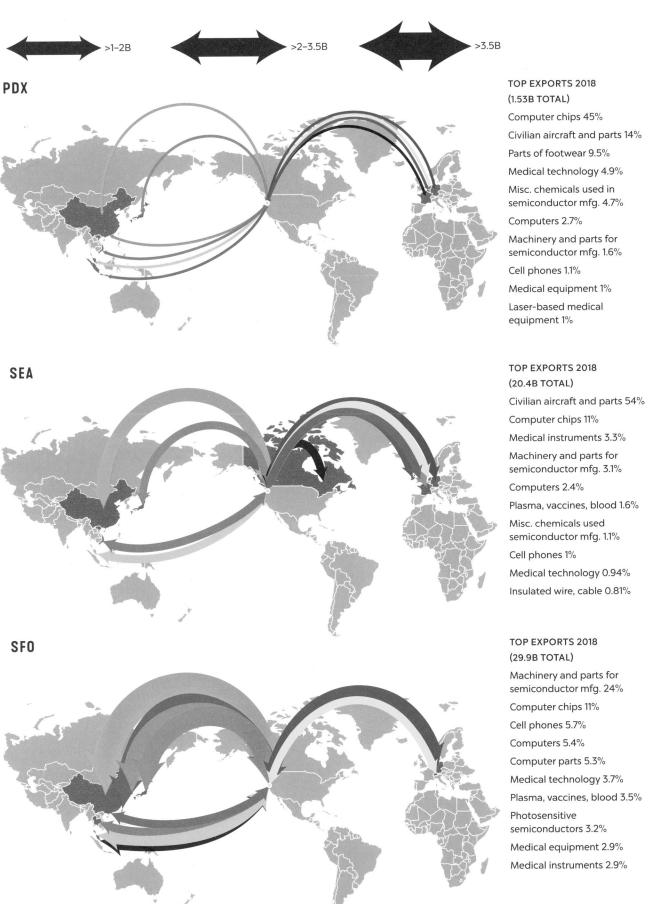

PDX

>1–2B   >2–3.5B   >3.5B

TOP EXPORTS 2018
(1.53B TOTAL)

Computer chips 45%

Civilian aircraft and parts 14%

Parts of footwear 9.5%

Medical technology 4.9%

Misc. chemicals used in
semiconductor mfg. 4.7%

Computers 2.7%

Machinery and parts for
semiconductor mfg. 1.6%

Cell phones 1.1%

Medical equipment 1%

Laser-based medical
equipment 1%

SEA

TOP EXPORTS 2018
(20.4B TOTAL)

Civilian aircraft and parts 54%

Computer chips 11%

Medical instruments 3.3%

Machinery and parts for
semiconductor mfg. 3.1%

Computers 2.4%

Plasma, vaccines, blood 1.6%

Misc. chemicals used
semiconductor mfg. 1.1%

Cell phones 1%

Medical technology 0.94%

Insulated wire, cable 0.81%

SFO

TOP EXPORTS 2018
(29.9B TOTAL)

Machinery and parts for
semiconductor mfg. 24%

Computer chips 11%

Cell phones 5.7%

Computers 5.4%

Computer parts 5.3%

Medical technology 3.7%

Plasma, vaccines, blood 3.5%

Photosensitive
semiconductors 3.2%

Medical equipment 2.9%

Medical instruments 2.9%

IS THERE GAS IN THE CAR?

Anyone traveling through a city by car will note that gasoline prices can vary widely, often seemingly without rhyme or reason. San Francisco, Portland, and Seattle are no exceptions. How do these cryptic gas prices get set?

Gas prices do not just reflect the cost of crude oil; there are other charges baked into each dollar spent at the pump. Turns out, though the oil industry as a whole is making big money, gas stations actually don't pull in high profits. Most gas station operators don't make their money from sales at the pump, but from retail goods, such as beer and cigarettes. Oil companies don't give vendors much of a discount, unless vendors buy large amounts, which most small stations cannot afford.

To stay competitive, many stations mark gas up only a few cents per gallon. Factors that might affect how much a gas station charges per gallon include location, quantity purchased, rent costs, and branding (brand name versus generic). State and local taxes also affect price. California's gas prices are the highest in the contiguous United States, in part because California's clean-gas laws are stricter than federal standards.

Often the reason gasoline costs more in some neighborhoods than in others is zone pricing. Gas pricing zones, also known as market areas, are a means for distributors to set a selling price to vendors based on region. The zone price is determined by an undisclosed recipe, often including traffic volume, local income levels, and local competition.

In the past, zones were mostly defined by city borders, but more recently, according to some independent vendors, zone borders have become like gerrymandered districts. Distribution companies can draw the boundaries of gas price zones however they want; two stations across the street from each other may be paying very different wholesale prices. Zone pricing also means that the station charging the highest prices might actually be making the lowest profit margin.

The data for these maps was collected during the first two weeks of June 2019. Across the cities, there are few gas stations in downtowns. In San Francisco, gas prices were the most expensive in South of Market, an area of highway on-ramps near the Bay Bridge. Prices were relatively high in the Mission, the Castro, and part of the city near Highway 101. Gas tended to cost less the further west you went except for an area outside of Land's End in the northwest corner of the city.

In Portland, the highest gas price occurred just outside of downtown, an area with access to I-5, as well as to several state and local routes. Gas prices remained relatively high throughout most of the west side of the city except for the northwesternmost station in the city. On the east side, the highest prices were also in areas near highway on-ramps.

Seattle's most expensive gas was in Madison Park. Gas was relatively expensive outside of downtown (where you can't buy gas) and in most of the southeast part of the city, including Rainier Valley. The pattern in the north was highly variable, as sometimes relatively expensive gas was located near relatively cheap gas.

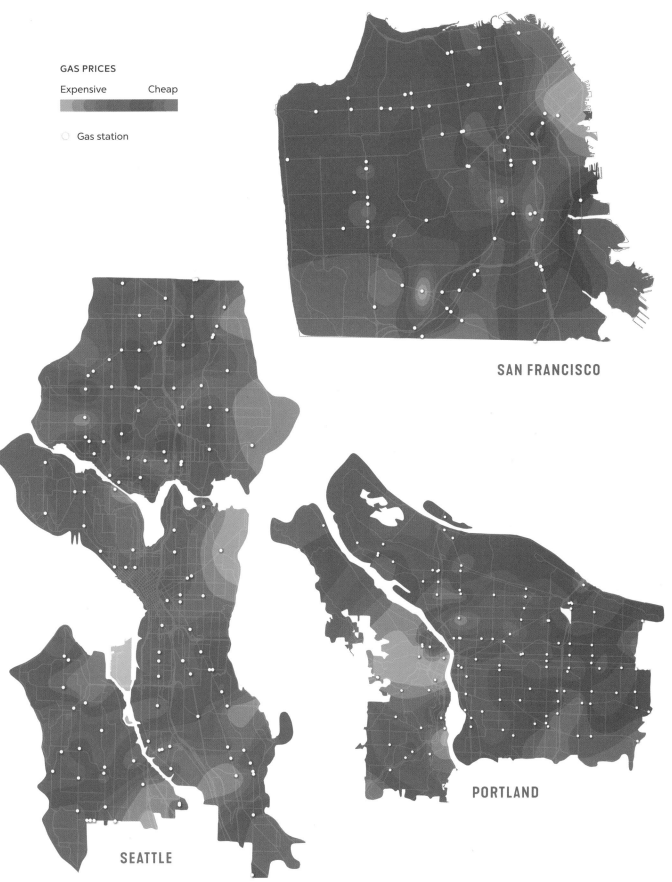

GAS PRICES

Expensive   Cheap

○ Gas station

SAN FRANCISCO

PORTLAND

SEATTLE

143

HOW GREEN IS MY ENERGY?

San Francisco, Portland, and Seattle all have green reputations. Part of being green involves using renewable energy sources and all three are ranked among the top ten green cities in the country. The demand for green energy in these cities exceeds what's available. Here we examine the sources of energy for each city. Calculating how much green energy a city uses is not as straightforward as city leaders would like us to imagine. There is often confusion between carbon-free energy, which includes nuclear power and hydropower from large dams, and renewable energy, which does not. Ever-changing state renewable energy requirements also drive the purchase of renewable energy from sources other than what utilities own.

In terms of green power generation (based on 2017 data), Oregon ranks fourth nationally, Washington ranks ninth, and California ranks tenth. The states that generate the most renewable power are Texas and Oklahoma, which generate massive amounts of wind power.

San Francisco aims for 100 percent renewable energy use and has an organization called CleanPowerSF to advance that goal. CleanPowerSF offers an energy option called SuperGreen that claims to eliminate the carbon footprint of your personal electricity supply with 100 percent renewable energy. However, the story of San Francisco's energy is tied up with California energy provider Pacific Gas and Electric (PG&E).

Although San Francisco obtains all of its energy through PG&E, the city calculates renewable energy differently than PG&E. PG&E buys rather than generates most of the renewable energy it provides. It has numerous hydroelectric units, but most are not classified as renewable because they are "large hydro," which according to the California legislature should not be counted as renewable because of the threat to salmon and other species of endangered fish. By contrast, Portland and Seattle consider all hydropower to be renewable. PG&E has plants powered by fossil fuel and a nuclear facility at Diablo Canyon. In other words, San Francisco claims to be greener than the company that delivers all of its power.

San Francisco leaders would prefer to outright own the means of power generation. In the beginning of 2019, PG&E filed for bankruptcy after being found liable for several California wildfires. This prompted the city of San Francisco to make a $2.5 billion bid to take over the power grid assets within city limits. The city claimed it could provide a safer, more updated power infrastructure at a lower cost. PG&E refused to sell. The utilities company exited bankruptcy in June 2020, a few weeks after making public its intention to move headquarters from San Francisco to Oakland.

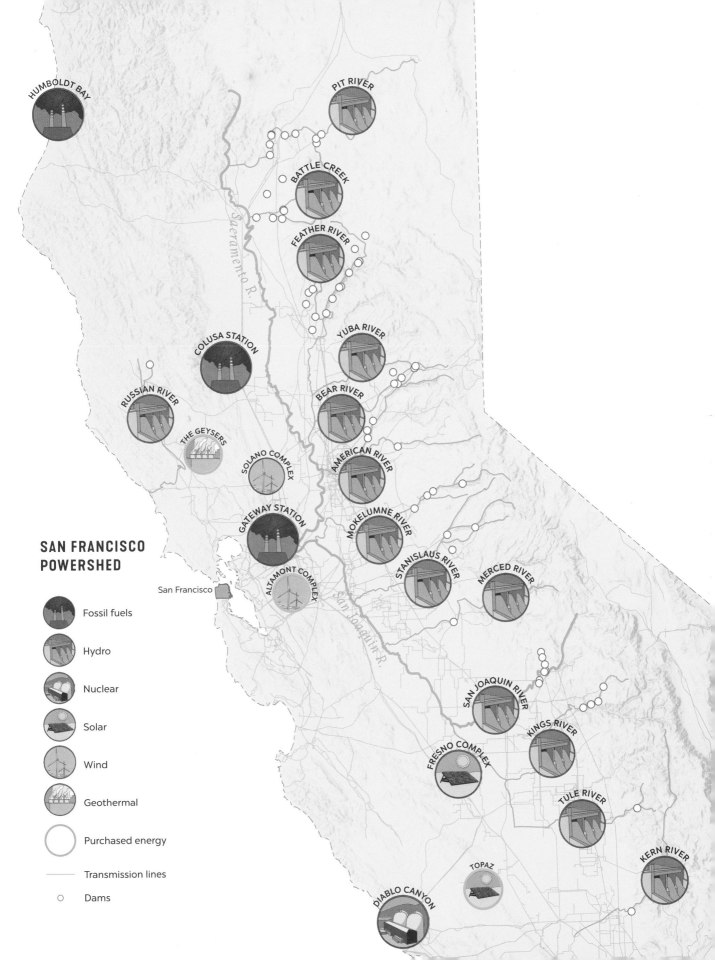

HUMBOLDT BAY

PIT RIVER

BATTLE CREEK

FEATHER RIVER

COLUSA STATION

YUBA RIVER

RUSSIAN RIVER

BEAR RIVER

THE GEYSERS

SOLANO COMPLEX

AMERICAN RIVER

GATEWAY STATION

MOKELUMNE RIVER

ALTAMONT COMPLEX

San Francisco

STANISLAUS RIVER

MERCED RIVER

Sacramento R.

San Joaquin R.

SAN JOAQUIN RIVER

KINGS RIVER

FRESNO COMPLEX

TULE RIVER

KERN RIVER

TOPAZ

DIABLO CANYON

SAN FRANCISCO POWERSHED

- Fossil fuels
- Hydro
- Nuclear
- Solar
- Wind
- Geothermal
- ◯ Purchased energy
- —— Transmission lines
- ○ Dams

More than 90 percent of Seattle's power is generated by hydropower. The next largest contributor to Seattle energy usage is nuclear, at 4 percent. Seattle uses very small amounts of nonrenewable sources including natural gas and coal, and their website is upfront about this fact. The city's Green Up program offers a program for businesses and residents to pay a little more each month to use carbon-neutral sources of energy. Seattle City Light serves a number of small nearby cities as well as the entire city of Seattle.

About half of Seattle's power generation comes from facilities they own, including the Boundary Dam on the Pend Oreille River and the Gorge, Diablo, and Ross hydroelectric plants and dams on the Skagit River. Seattle buys the other half of their energy, the majority of which is hydroelectric power that comes from the Bonneville Power Administration (BPA). The majority of BPA dams are "large hydro" plants.

Portland has set out to generate or purchase 100 percent of energy for city operations from renewable sources. The story of renewable energy in Portland has more to do with wind power. Portland General Electric (PGE) is Portland's public power utility. Similar to Seattle, there is an option to pay a bit more on your monthly bill to increase the share of renewable power used.

Portland was the last of the three cities with significant coal power usage until the Boardman coal facility closed late 2020. At that time more than 40 percent of Portland's energy came from a mix of fossil fuels.

Portland General Electric serves the entire Metro Area (the part within Oregon) except, interestingly, a portion of Northeast Portland and parts of downtown, which are served by PacifiCorp, once part of the Enron energy empire.

SEATTLE AND PORTLAND POWERSHEDS

- Fossil fuels
- Hydro
- Large hydro
- Wind
- Purchased energy
- Transmission lines
- Seattle City Light dams
- Portland General Electric dams
- Bonneville Power Administration dams

CLATSKANIE CO

WILLAMETTE

SKAGIT RIVER

Gorge
Diablo
Ross

Boundary

PEND OREILLE RIVER

SOUTH FORK TOLT RIVER

Seattle

Tolt
River

Wells
Chief
Joseph

Grand
Coulee

Rocky Reach

Rock Island

COLUMBIA RIVER

Wanapum

Priest Rapids

Hanford

Lower
Monumental

Little Goose

SNAKE RIVER

Lower
Granite

Ice Harbor

TUCANNON

McNary

STATELINE

WASHINGTON

OREGON

Bonneville

John Day

The Dalles

BIGELOW CANYON

BOARDMAN COMPLEX

lamette Falls
River Mill
CLACKAMAS RIVER

aday

arriet Lake

Round Butte
Pelton

DESCHUTES RIVER

147

LIVES AND LIVELIHOODS AT RISK: COVID-19

COVID-19 is a public health disaster that quickly became an economic crisis too. Some lost loved ones, some lost their jobs, and others lost both. Few of us saw this coming in early January of 2020. For many people, the transition from normal workday or school day to something quite different occurred overnight. The Upper Left stands out as the place where COVID-19 was first reported in the United States and for having early hot spots of recorded cases but then having lower reported cases relative to many other parts of the country ten months later.

The coronavirus's impact in the Upper Left first hit the Seattle area, then the Bay Area, and subsequently around Portland. On January 20, the first documented case of COVID-19 in the United States was recorded in Snohomish County, Washington, just north of Seattle. The patient, who recovered, was a thirty-five-year-old US citizen who had recently returned from a trip to Wuhan, China. Upper Left connections to the coronavirus may predate this case, as a November 2020 study found coronavirus antibodies present in American Red Cross blood samples from California, Oregon, and Washington that date back to December 2019.

Although the first US death from COVID-19 was reported near Seattle on February 29, an autopsy report released in late April determined that a fifty-seven-year-old woman in San Jose, California, died from the disease on February 6. The woman is believed to have contracted the coronavirus through the local community. The areas around Seattle and San Francisco went on to became initial centers of recorded COVID-19 cases. Seattle released its first report on local COVID-19 cases in February. San Francisco began reporting COVID-19 statistics in mid-March. Portland's first report on COVID-19 cases came in May.

On March 19, California became the first state to issue a stay-at-home order for nonessential workers and nonessential travel. Oregon and Washington issued similar stay-at-home orders on March 23. This immediately triggered a massive unemployment crisis. National employment rates soon exceeded those of the 2008 recession.

In April, the governors of California, Oregon, and Washington announced a pact for reopening their economies and controlling the spread of COVID-19 by building out a tripartite West Coast framework. As the months went by, the West Coast governors moved slowly in reopening relative to many other states in the country.

The top map represents reported cases per capita through November 2020. The Upper Left as a region is within the lowest category of active cases, except for most of the Bay Area, which is in the second-lowest category. An October 2020 Seattle mayor's office study found that of the country's thirty largest US cities, Seattle itself had the lowest rate of COVID-19 cases per capita. The same study ranked Portland (Multnomah County, really) as having the second-lowest rate.

The bottom map represents the cumulative deaths per capita through November 2020. Although the Bay Area looks to be close to the national average, much of the Upper Left has fewer deaths per population relative to many places across the country.

CUMULATIVE COVID-19 CASES PER CAPITA
NOVEMBER 2020

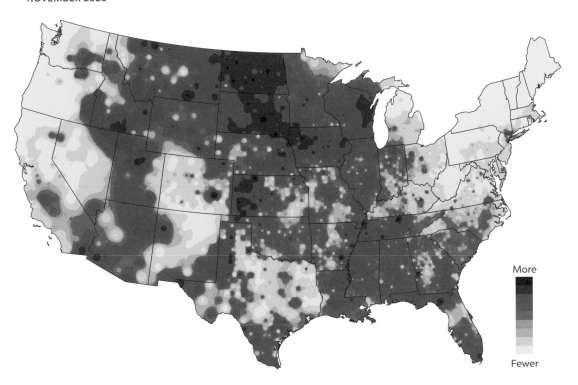

More

Fewer

CUMULATIVE COVID-19 DEATHS PER CAPITA
NOVEMBER 2020

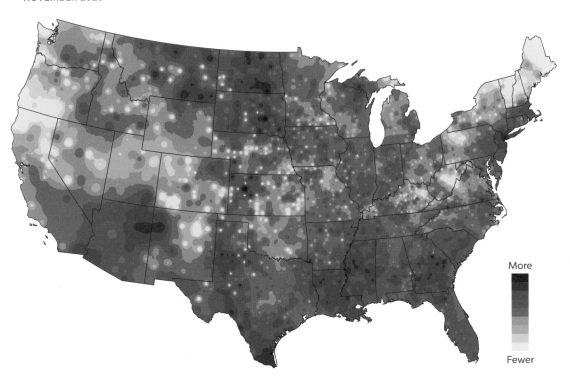

More

Fewer

We looked at the pattern of COVID-19 cases per capita by zip code, which was the best data we found that could be compared across the three cities. We also looked at dates in April and June and found the patterns on those dates to be largely consistent with those we mapped on these pages for November 2020.

It is important to understand that these maps represent the home locations of people who have tested positive for the virus. The maps have limitations. They do not map the actual incidence of COVID-19 because so many people have not been tested and rates of testing vary widely. They also do not map where the virus was contracted.

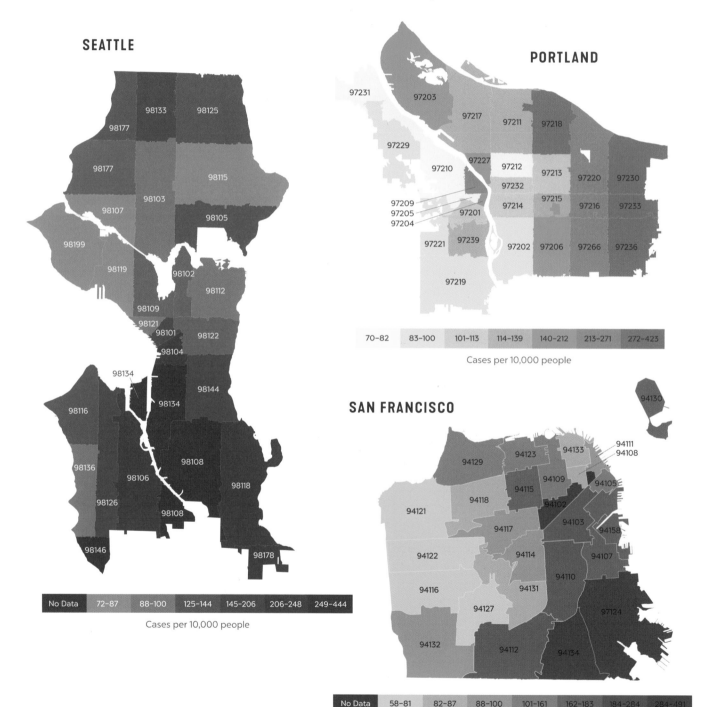

SEATTLE

PORTLAND

70–82	83–100	101–113	114–139	140–212	213–271	272–423

Cases per 10,000 people

SAN FRANCISCO

No Data	72–87	88–100	125–144	145–206	206–248	249–444

Cases per 10,000 people

No Data	58–81	82–87	88–100	101–161	162–183	184–284	284–491

Cases per 10,000 people

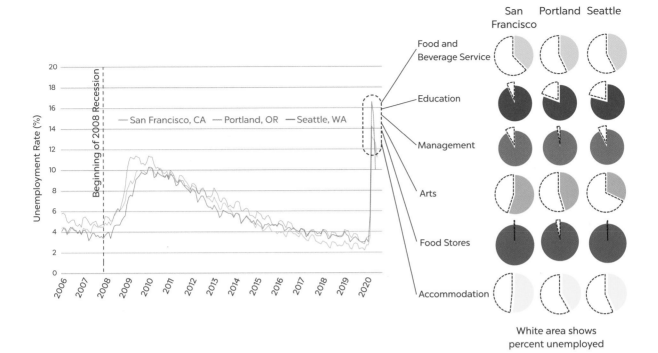

San Francisco · Portland · Seattle

Food and Beverage Service

Education

Management

Arts

Food Stores

Accommodation

White area shows percent unemployed

Patterns of COVID-19 cases have something to do with where essential workers live—the geographies of who are the most exposed to contact with others. What we can see from these maps is that the highest per capita rates of COVID-19 cases are in some of the poorest and most nonwhite parts of each city. The crisis made clear the great importance of often-underpaid essential workers (including nurses, grocery store workers, cooks, delivery people, and trash collectors) and the daily risks they take to do their jobs. Other heroes of the pandemic include the secondary school teachers who had to immediately learn to instruct large classes of children remotely.

We also compared the rates of unemployment from April 1, 2019 to April 1, 2020 when job loss related to COVID-19 was beginning to peak. These graphs show that some industries were hit much harder than others. Workers in accommodations, food service, and the arts experienced the most dramatic rates of job loss. Much less affected were general merchandise, professional services, and management.

Cities began to reconfigure parts of the urban landscape, closing streets to vehicular traffic and expanding sidewalk seating for restaurants, bars, and cafés. Another indicator of how much life changed is the stark drop in public transportation ridership the cities experienced.

BART ridership to San Francisco — April 2019 / April 2020

TriMet ridership in Portland Metro Area — April 2019 / April 2020

King County Transit ridership in Seattle Metro Area — April 2019 / April 2020

NONSTOP HUB HOP

Airports are often seen as the gateway to any city in the world. And for good reason! Most visitors begin and end their trips at the airport. According to the World Bank, there were nearly four billion passenger trips across the globe in 2017. People from the United States took nearly 850 million of those trips.

As any frequent flier will tell you, not all airports are created equal nor are all flights. Depending on the size of the airport, the size of the city, the business within the city, and many other factors, your local airport could have no direct flights or hundreds. International flights, in particular, are often seen as a symbol that a city has "made it." Where and when you can fly around the globe without stopping somewhere else first also matters.

In looking at the three cities in the Upper Left pre-COVID, there is a pretty strong relationship between the size of the city or metropolitan area and the number of places one can fly to directly. Portland, the smallest of the bunch, has the fewest daily nonstop international flights, with only eleven. Seattle, a bigger city, has twenty-five nonstop international flights, mostly to Asia. Between San Francisco International Airport (SFO) and Oakland International Airport (OAK), San Franciscans and other Bay Area residents can book nonstop flights to fifty-nine international destinations.

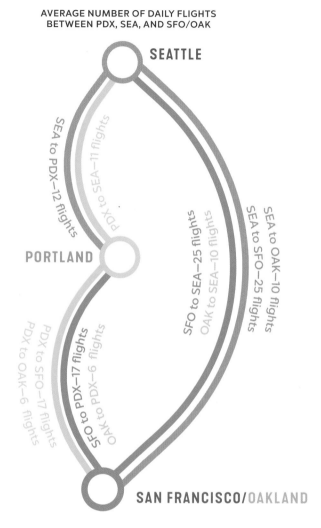

AVERAGE NUMBER OF DAILY FLIGHTS BETWEEN PDX, SEA, AND SFO/OAK

SEATTLE

SEA to PDX—12 flights
PDX to SEA—11 flights

PORTLAND

SFO to SEA—25 flights
OAK to SEA—10 flights

SEA to OAK—10 flights
SEA to SFO—25 flights

PDX to SFO—17 flights
PDX to OAK—6 flights

SFO to PDX—17 flights
OAK to PDX—6 flights

SAN FRANCISCO/OAKLAND

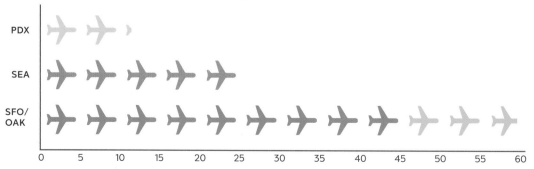

NUMBER OF INTERNATIONAL DESTINATIONS

PDX

SEA

SFO/OAK

| 0 | 5 | 10 | 15 | 20 | 25 | 30 | 35 | 40 | 45 | 50 | 55 | 60 |

SEATTLE-TACOMA INTERNATIONAL AIRPORT

PORTLAND INTERNATIONAL AIRPORT

SAN FRANCISCO/OAKLAND INTERNATIONAL AIRPORTS

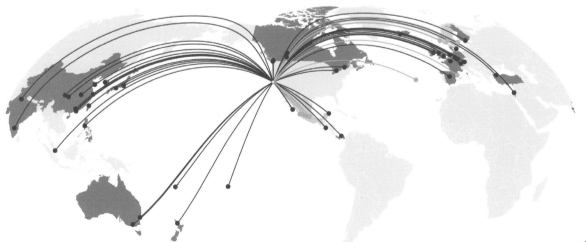

ANALOG: BACKLASH TO DIGITAL

The technology we use for work and enjoyment has changed radically over the last thirty years. In the 1980s, analog technologies were dominant, though at the time we didn't call them analog technologies—we just thought of them as the machines of contemporary life: typewriters, cameras, record players, and pianos. They had been around for a good while and remained similar to versions from a hundred years prior.

How did business get done in the 1980s? Typewriters. Lots of typewriters. Commercial typewriters have been available since the 1870s. Standard in offices by the 1880s, the typewriter remained one of the most important machines for conducting business a hundred years later. This required lots of typewriter stores for sales and repairs.

How did people record and view images in the 1980s? The answer is film cameras, which are innovations of the eighteen hundreds. In 1889, George Eastman began manufacturing celluloid film for his new camera—the Kodak. A century on, film was still the primary means by which people recorded, archived, and viewed images. Camera stores were still common in the 1980s; that decade saw the rapid development of one-hour photo facilities operated out of pharmacies, supermarkets, and malls.

How did people enjoy music in the 1980s? Although this would be the decade vinyl record sales fell behind cassette tape sales, record players and record stores still figured prominently. Even when shops sold cassette tapes and later CDs, these places were still usually called record stores. The phonograph (a.k.a. gramophone, a.k.a. record player) was invented in 1877, becoming commercially available in the 1880s and still going strong in the 1980s.

How about making your own music in the 1980s? The answer for many was the piano. The first piano is thought to date to the late 1690s, but the era of the modern piano began about one hundred years later. By the 1880s, upright pianos had become widely adopted as space-saving solutions for the smaller living spaces of the industrializing city. Fast-forward to the 1980s and electric keyboards are commercially available, but still stood in the shadows of the piano.

To better understand just how widespread these technologies were in the 1980s, we decided to map typewriter stores, film developing outlets, record shops, and piano stores for the year 1987 in San Francisco, Portland, and Seattle. How did we get the data? Phone books from 1987.

Things have changed. In 1987, there were forty-six typewriter shops in San Francisco, forty-two in Portland, and thirty-four in Seattle. Today, Portland has two typewriter shops and the other two cities have none.

In 1987, San Francisco had seventy-four record stores. In 2019, it had just twenty-one. Seattle went from forty-four to seventeen. In Portland, twenty-six piano shops dwindled down to nine. Film developing is harder to track today, but our research suggests that, although digital prints are easy to get, finding places to develop film is difficult. There are other surviving analog businesses that soldier on in the digital world including bookstores, craft stores, and stationery stores.

In the following pages, the maps depicting analog businesses (from 1987) were created by hand by one of our artists. The data was organized and mapped on a computer, then interpreted into three different mediums: needlepoint, drawing, and printmaking. Each map is captioned with thoughts from their creator.

PORTLAND

Embroidery offered a different surface material to work with, more detailed, yet simplistic. The icons representing each location are all the same shape, only distinguishable by their different colors.

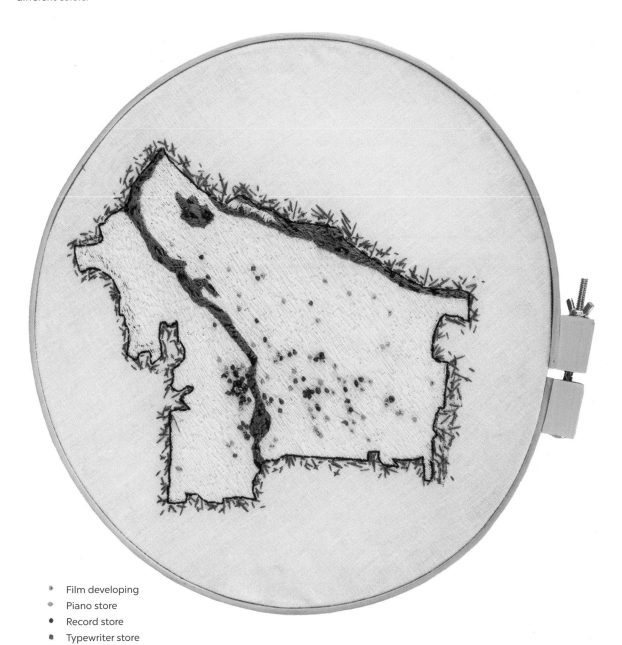

- Film developing
- Piano store
- Record store
- Typewriter store

SAN FRANCISCO

San Francisco's map borrowed the aesthetic of ancient seafaring charts with the added impudence of stickers. Stickers were created and used for their contemporary functions as self- and place-branding tags.

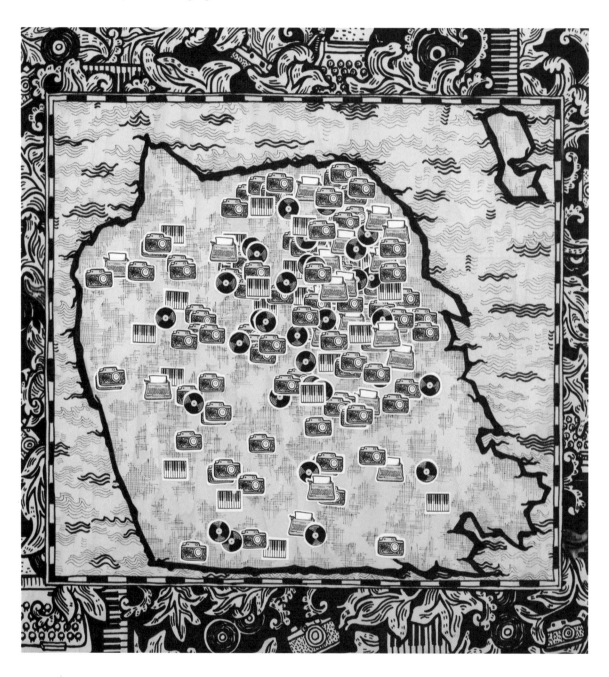

 Film developing   Piano store   Record store    Typewriter store

SEATTLE

The Seattle map combined stark black block-printed architectural images on layers of paper with the lightness of colorful buttons, carefully stitched to overlap and contrast with each other and their shared environment.

Film developing

Piano store

Record store

Typewriter store

ANALOG IN A
DIGITAL WORLD

These maps show the remaining analog shops in our increasingly digital world. Their numbers have dwindled and their spread is less extensive. Still, enough people are sufficiently old school to keep these businesses going. Dedication to vinyl records is apparent. A surprising number of piano stores continue to operate. Photo shops are harder to find than before but are still around. Typewriter fans have it tough. Across the three cities, we found only two typewriter shops, plus a camera store that works on typewriters—all in Portland.

Photo developing store
Piano store
Record store
Typewriter store

10000110 11110100
11100100 10011010
00101100 01111011
00100100 01000111
10001000 11110011 00001000
01011001 01101111 01001010
10000000 01011100 00000110
10010100 10100011 01110101 11110101
10111111 10101000 10100101 00111100
11011001 00111011 11001010 10101010
01001000 00010101 00101110 10010001
10000100 10000101 10110010 10010000
11010111 01010111 11101001
00000000
01010010 00101011
10001000 11001100
00110000 10101111
01000111 00100001
01000001 11011101
01000111 11011001
10011001
01010111
10101110
00101110
11110010 11100011 00110101
10001101 00001101 01001111
10010000 10000011 00110110
11011110 11010110 11101011 10101101
01011111 11110001 10101101 10001100
10110011 10010111 11010100 01110011
10101101 01100100 01000111
00111011 00100001
00100010 00000011 00111111
01011000 10000100 10101101
100100 00
01111 010
010 01111
10

158

01
0100
01000001
01000011 01010100
00100000 01001001 0001010
01010000 01010010 01001111 01001100 01001111
01000111 01010101 01000101 0001010 01010100
01110111 01101111 00100000 01101000 01101111
01110101 01110011 01100101 01101000 01101111
01101100 01100100 01110011 00101100 00100000
01100010 01101111 01110100 01101000 00100000 01100001
01101100 01101001 01101011 01100101 00100000 01101001
01101110 00100000 01100100 01101001 01100111
01101110 01101001 01110100 01111001 00101100
0001010 01001001 01101110 00100000 01100110
01100001 01101001 01110010 00100000 01010110
01100101 01110010 01101111 01101110 01100001
00101100 00100000 01110111 01101000 01100101
01110010 01100101 00100000 01110111 01100101
00100000 01101100 01100001 01111001 00100000
01101111 01110101 01110010 00100000 01110011
01100011 01100101   01101110
01100101
00101100

0111   0101 01110100
00100000 01110010 01100101
01101100 01100001 01110100
01101001 01101111 01101110 01110011
00100000 01110100 01101111 00100000 01110100
01101000 01100101 00100000 01101001 01101110
01100100 01101001 01110110 01101001 01100100
01110101 01100001 01101100 00100000 01110100
01101000 01101001 01101110 01101011 01100101
01110010 00101110 00100000 01010100 01101000
01100101 01111001 00100000 01100001 01110010
01100101 00100000 01101101 01100101 01100001
01110011 01110101 01110010 01100101 01100100
00100000 01100010 01111001 00100000 01101000
01101001 01110011 00100000 01110111 01101001
01101100 01101100 01101001 01101110 01100111 01101110
01100101 01110011 01110011 00100000 01110100
01101111 00100000 01100001 01100011 01110100
00101110

GOTTA GO TO WORK

These maps explore the directionality of commuting within San Francisco, Portland, and Seattle. The arrows indicate the average direction of the commute from that area. The darker the arrow, the higher the number of jobs in an area; the longer the arrow, the longer the commute distance. The red lines indicate some of the main destinations— the densest employment areas. In some cases, there is one company employing many people. In other cases, there are high-rise office buildings that hold many companies that collectively employ many people.

The sheer density of jobs in each downtown area stands out in each city. That commuting downtown is often hairy is also evident. In each city, public transportation also tends to serve downtowns better than any other area (try getting north–south on a bus in most parts of Portland). The fact so many people head downtown from so far away also suggests something about the lack of affordable housing in areas where most jobs are located.

SAN FRANCISCO

San Francisco's downtown is the center of a band of employment that reaches into Berkeley and Oakland in the East Bay and stretches south through Silicon Valley (which is probably bigger than most people think). Because of this, the flow of arrows in the area is less focused on downtown compared to Portland or Seattle. San Francisco International Airport operates within the southern employment band, and the Oakland port and airport are both within the employment concentration in the East Bay. The concentration of jobs in San Jose is a good reminder that the city has more than a million people.

The Bay Area traffic situation is notoriously horrible. Getting to San Francisco from outside the city by car is going to be brutal during anytime close to business hours. There is no gaming the Bay Bridge—you are probably going to wait in traffic. The map illustrates that many people commute to the city from as far away as the northern part of California's central valley.

Among large US cities, San Francisco has the third-largest percentage of commuters who walk, at 10.2 percent. San Francisco also has the fourth-highest rate of bicycle commuters with 3.1 percent.

Among the forty-some Fortune 500 companies located in San Francisco are Apple, Charles Schwab, Chevron, Electronic Arts, Gap, Google, Hewlett-Packard, Intel, Oracle, PG&E, Visa, and Wells Fargo.

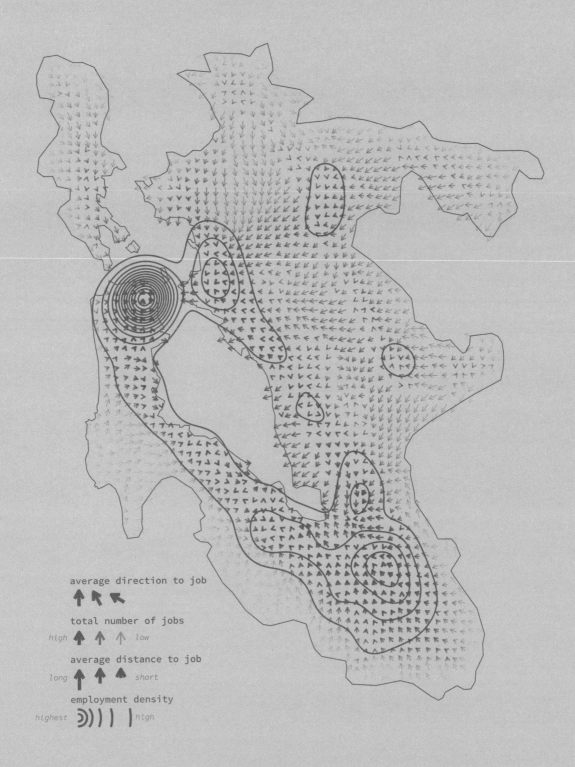

average direction to job

total number of jobs

high low

average distance to job

long short

employment density

highest))) high

PORTLAND

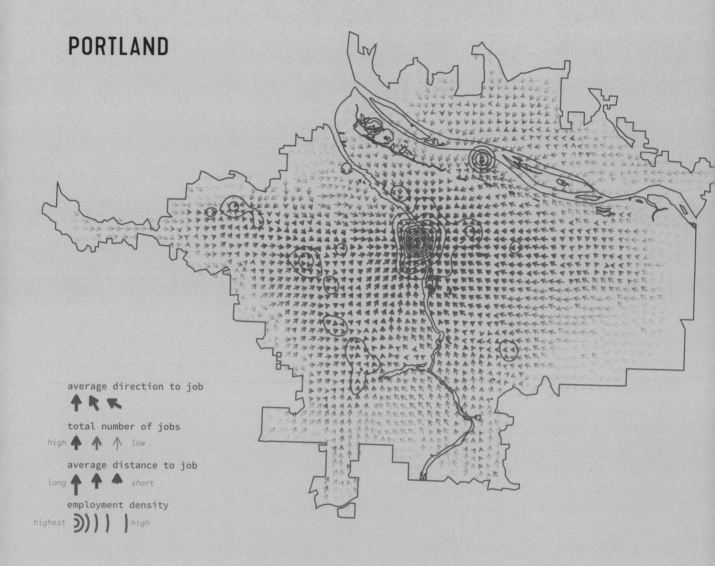

average direction to job

total number of jobs
high / low

average distance to job
long / short

employment density
highest / high

Throughout much of the metro region, the main flow of the commute is headed to downtown Portland. This is less true on the west side of the metro area where major employers Nike and Intel each draw thousands of commuters daily. The density of employment from downtown comes from the multiple businesses operating in tall buildings. Portland State University is one of the destinations that employs thousands of people. The concentric circles northwest of downtown are the Portland International Airport.

The driving commute in Portland is comparatively better than San Francisco and Seattle,

but compared to Portland of just five years ago, it's terrible.

Of the seventy largest cities in the United States, Portland has the highest rate of bicycle commuters with 6.3 percent. A similar percentage of people commute by walking.

The Portland area has one solitary Fortune 500 company in Nike. Up until 2015, it used to have two until Precision Castparts Corporation was sold to Warren Buffett's Berkshire Hathaway (ranked number four in 2019's Fortune 500) and for that reason no longer shows up on the list.

SEATTLE

Downtown Seattle looks like a target among the rings of employment that surround it. More so than the other two cities, commutes in Seattle are often highly restricted by water. Across Lake Washington to the east, Redmond's employment giant Microsoft jumps out. The arrows flow to downtown but also to Redmond. South of both Seattle and Redmond is SeaTac, the city that hosts the airport of the same name. The ring at the southwest end corner of the map is the Port of Tacoma. Boeing's Everett facility is the red circle in the north of the metro area.

Seattle traffic is pretty awful, by any measure. All workday long, I-5 is a parking lot and in Seattle, accidentally getting on a highway is pretty easy to do. If you work north or south of the city, leaving before six in the morning is your only chance of getting on the highway without waiting in a creeping line of slow-moving commuters.

Of the seventy largest cities in the United States, Seattle has the fifth-highest percentage of walking commuters with 9.3 percent and the sixth-highest percentage of bicycle commuters with 2.8 percent.

The ten Fortune 500 companies in the Seattle metro area are Alaska Airlines, Amazon, Costco, Expedia Group, Expeditors International of Washington, Microsoft, Nordstrom, Paccar, Starbucks, and Weyerhaeuser.

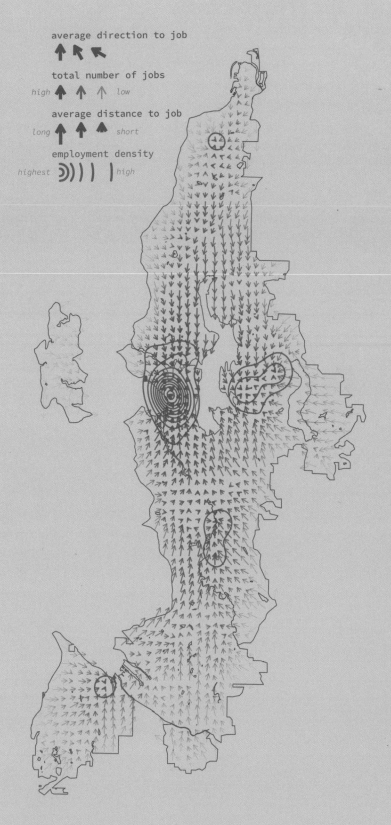

average direction to job

total number of jobs

high — low

average distance to job

long — short

employment density

highest — high

163

IN THE WEEDS

Upper Left cities have a bit of a reputation for being pot-friendly. San Francisco's history is peppered with subcultures known for smoking weed, from the jazz musicians and beat poets of the 1950s to the hippies who flocked to San Francisco during the Summer of Love in 1967. The Pacific Northwest developed a reputation for Mary Jane of its own—after all, that's where the Grateful Dead liked to hang out in their spare time and where Ken Kesey settled down (after getting busted for possession). Pot smoke lingered over the grunge scene of 1990s Seattle. Portland holds a special place in the country's popular imagination as a city where a lot of grass gets smoked.

Before there was recreational weed, there was medical marijuana. California was the first in the country to legalize medical marijuana (in 1996). Washington and Oregon were close behind, adopting their own medical marijuana laws two years later. They were three of the first four states to permit medical marijuana ('sup Alaska?). Washington, alongside Colorado, legalized recreational pot before other states, so Seattle got legal recreational weed first among Upper Left cities.

SAN FRANCISCO

WEED DELIVERY SERVICE

WEED DISPENSARY

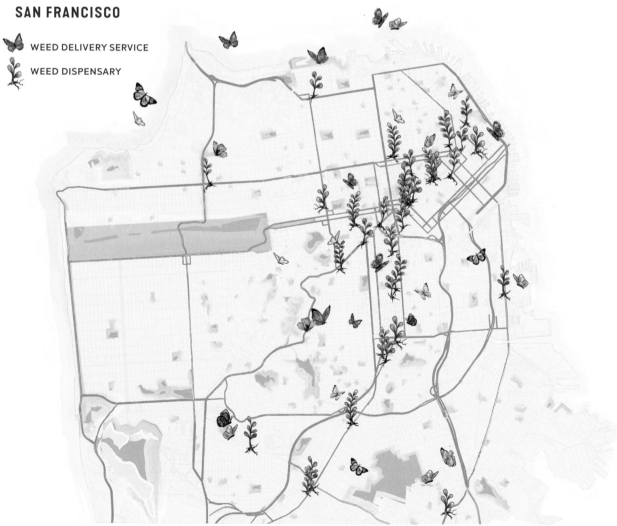

Recreational cannabis was legalized in Washington in 2012, Oregon in 2014, and California in 2016, cementing the stoner haven status of Upper Left cities. Each city has a top ten list of weed tours—a luxury afforded only to cities with thriving cannabis tourism industries. San Francisco has the second-largest number of weed shops in California (after Los Angeles) despite having the fourth-largest population in the state. Circa 2019, Oregon had grown so much ganja that supply was around ten times demand, resulting in retail prices that rivaled those on the street.

Legalizing pot is not just a cultural phenomenon but an economic one. Marijuana taxes are no small figures. Tax revenues for cannabis in 2018 were $319 million in Washington, $300 million in California, and $94.4 million in Oregon. San Francisco's tax rate on marijuana is 23.5 percent, Portland 20 percent, and Seattle a whopping 47 percent (37 percent of that comes from a state levy).

After legalization, brick-and-mortar marijuana shops sprouted up across Upper Left cities. On the maps, look out both for weeds (pot shops) and butterflies (delivery-only businesses). The cities vary in how much they embrace a weedy garden. San Francisco has a more manicured look, as butterflies flit from place to place. Portland lets its weeds run wild. Seattle is a garden without butterflies.

In 2019, when we collected our data, San Francisco and Seattle each had sixty active licenses for retail marijuana. Portland had more retail marijuana shops than the other two cities combined with 164. Considering that Portland also has a smaller population than either San Francisco or Seattle, Portland may well be the Upper Left capital of legal weed.

The weed delivery scene varies widely in these cities. Almost half of San Francisco's retail weed operations are delivery only—leaving only thirty-one brick-and-mortar shops citywide. In Portland, there are six delivery-only marijuana businesses, but many retail shops also offer delivery. In Seattle, weed delivery is illegal, but of course you can get

SEATTLE

🌿 WEED DISPENSARY

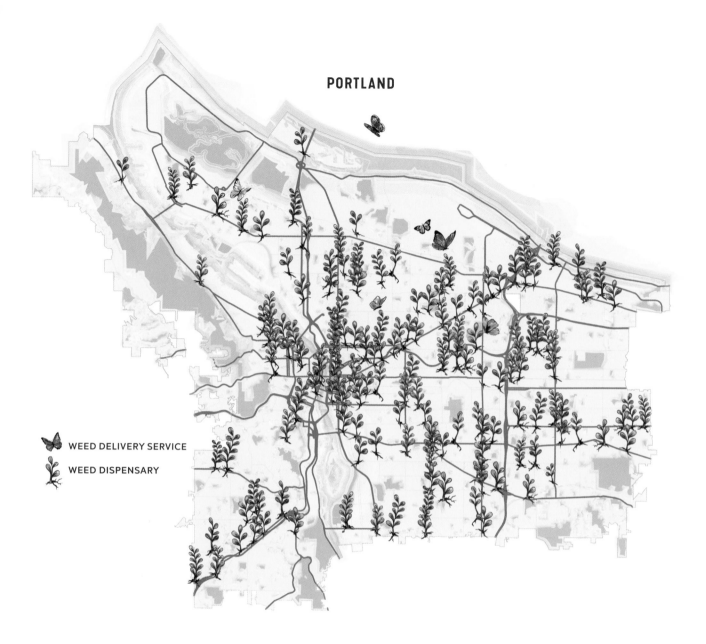

WEED DELIVERY SERVICE

WEED DISPENSARY

pot delivered illegally (just like you could before legalization).

Nearly all of San Francisco's shops are east of Golden Gate Park. Several stores are on or south of Market and a few more are in the Mission. Most neighborhoods don't have a shop. Curiously, the Haight is nearly devoid of shops (but a wide selection of glass bongs shaped like mushrooms and wizards is still readily available).

Not only does Portland have a much greater number of retail shops, but it has a much wider spread. The main exceptions to an otherwise weedy city are the West Hills and the southeast-ernmost part of the city. These areas seem to be relative weed deserts (or orderly gardens). A string of shops along a diagonal in Northeast Portland span the length of Sandy Boulevard, which is sometimes called the Green Mile.

Seattle's situation is more like San Francisco's than Portland's. Seattle's shops are bunched in a number of areas, including Wallingford, Fremont, Capitol Hill, and the Industrial District. Many densely populated neighborhoods have no shops at all. Seattle's weed belt runs north to south.

When it comes to shop names, groan-inducing marijuana references abound. Each city seems to place a premium on puns and less-than-subtle one-word references to stoner culture. Seattle boasts the Joint, the Reef, OZ., and Canna Republic. Portland counters with Jayne, Bloom, Terpene Station, and Gram Central Station. San Francisco hosts SPARC, Bud, NUG, and Flower to the People.

Time will tell how these garden maps will change. Will the cities become overrun with weeds and butterflies? Will the federal government come in with weed whackers and butterfly nets? Or will the sprouts and their winged pollinator friends eventually settle out only in certain neighborhoods?

The national map on this page shows the legal status of cannabis in the fifty states. The Upper Left, as part of the West Coast, forms a solid block of legal use punctured by Idaho. Three of the New England states and New Jersey are starting to form an East Coast legal use block (what's the holdup, Connecticut?). Some states have curious arrangements such as Nebraska, where cannabis is illegal but decriminalized. Many of the southern states allow only highly restricted medical use, yet two of them, North Carolina and Mississippi, decriminalized cannabis. In 2015, legal cannabis came to Washington, DC, where you can possess up to two ounces and grow up to six plants at home on city land but can't have or use any weed on the 20 percent of the city that is controlled by the feds.

MARIJUANA LEGALIZATION STATUS

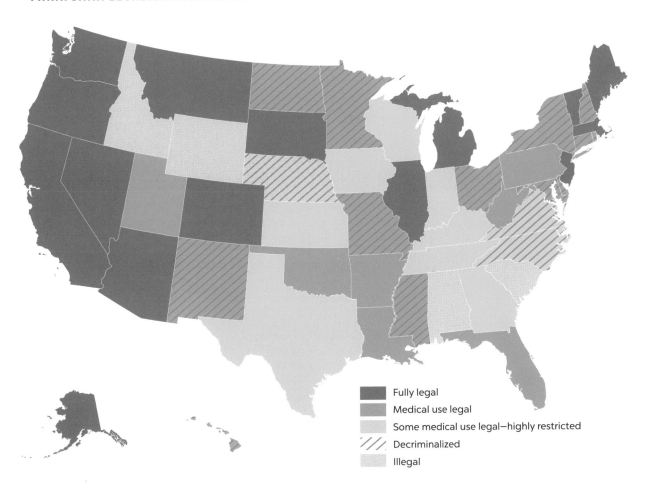

- Fully legal
- Medical use legal
- Some medical use legal—highly restricted
- Decriminalized
- Illegal

V. POPULAR CULTURE

POPULAR CULTURE IS THE WAY we live every day—the food we eat, the games we play, the technology we use, and the places we go for fun. Sometimes dismissed as frivolous and unworthy of attention, popular culture is what many people care most about. Popular culture and subcultures intermingle; what was once subculture—say, graffiti art—enters the popular culture; what was once popular culture—we're looking at you, typewriters—becomes a subculture.

Subcultures give cities character. Some subcultures are widespread in the Upper Left, such as foodies, gamers, and sports fans. San Francisco, Portland, and Seattle have reputations for celebrating quirky and off-beat subcultures, such as artisanal whiskey production or Roller Derby.

Each of these cities have had their moment in the pop culture sun and continue to be associated with their most notable cultural moments—San Francisco in the late 1960s, Seattle in the early 1990s, and Portland in the 2010s. San Francisco's 1960s' identity still gives it a lingering hippie and countercultural veneer. There is still a latent grunge association in Seattle, thirty years after its high point in the '90s. Portland, with its reputation for hipsters and makers, is still riding its recent cultural wave.

In this chapter, we examine some aspects of popular culture shared by San Francisco, Portland, and Seattle.

THE THRILL IS ALMOST GONE

Amusement parks trace back to sixteenth-century European pleasure gardens but have been reinvented over time. Designed to entertain and amuse, amusement parks include many forms of entertainment. In the United States, Coney Island is legendary. In fact, Coney Island is a large neighborhood that at its height hosted three amusement parks—Steeplechase Park, Luna Park, and Dreamland. The country's first roller coaster, the Switchback Railway, debuted on Coney Island in 1884.

The idea of dedicated spaces for amusement coincided with the emergence of electric trolley lines in American cities. Trolley companies realized that if they gave people a place to go on the weekends, profits

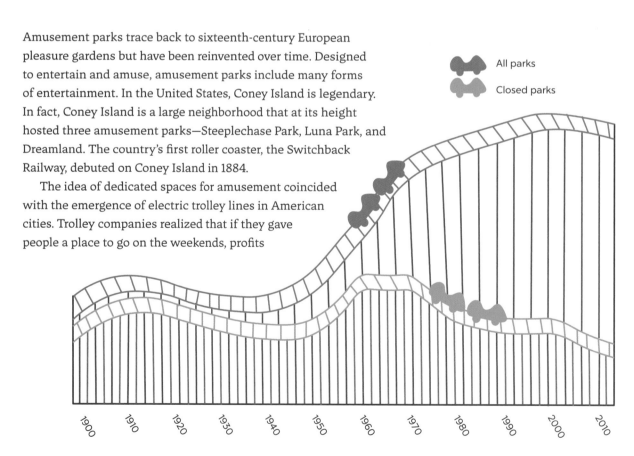

All parks

Closed parks

1900 1910 1920 1930 1940 1950 1960 1970 1980 1990 2000 2010 would go up. The scene was set for amusement parks to thrive and by 1910, there were more than two hundred amusement parks all over the country. Amusement parks took a big hit during the Great Depression and WWII, but enjoyed a comeback in the 1950s when Disneyland first opened, and again in the 1990s.

Many culturally significant amusement parks have closed their doors. Some parks were open for brief periods, such as Seattle's own Luna Park. Others provided decades of entertainment, such as Playland on Bitter Lake, which operated for just over three decades.

The only remaining amusement park within the city limits of the three cities is Portland's Oaks Amusement Park. Located on the east bank of the Willamette River near the Sellwood neighborhood, Oaks Park has been operating continuously since 1905. Impressively, the park also hosts a roller-skating rink that has operated for just as long. Oaks Park premiered its newest roller coaster, the Adrenaline Peak, in 2018.

Of the three cities, Portland also holds the record for the shortest running amusement park, Lotus Isle, which opened up to compete with the nearby Jantzen Beach amusement park. For the two years it was open, it was plagued with disaster. The ballroom burned down one summer night in 1931, a plane crashed into an artificial mountain on the scenic railway the same year, and the day after the first season ended in 1930 the president of Lotus Isle committed suicide.

After Disneyland opened in the 1950s, many other parks subsequently opened throughout the state. Although thirty-four amusement parks (many now closed) have operated in Northern California, only one amusement park ever existed in the city limits of San Francisco—Playland-at-the-Beach.

The current trend is to open parks in less-populated areas, rather than in cities. With increasing populations and a culture that promotes the use of personal transportation, placing amusement parks outside of city limits may make economic

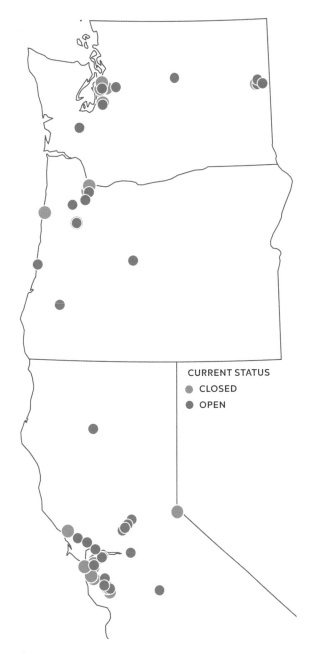

CURRENT STATUS
- CLOSED
- OPEN

sense, but distances people from them. The loss of the urban amusement park has also been a loss of a cultural pastime for urbanites.

For the amusement parks that have closed their doors in the Upper Left, the temporal rise and fall follow different patterns. Most closed their doors decades ago, with barely a trace left today of their existence. Many still live on in the memories of people who visited these amusement parks in their youth.

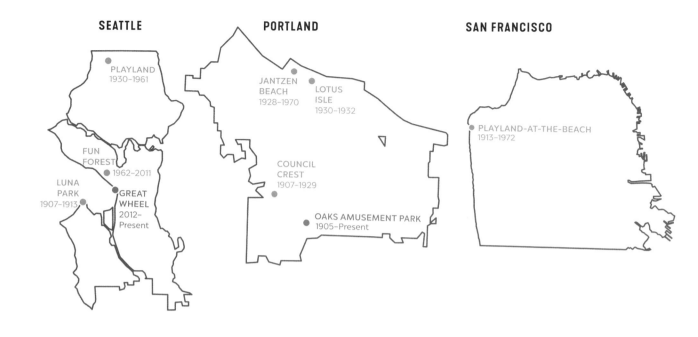

SEATTLE

PLAYLAND
1930–1961

FUN FOREST
1962–2011

LUNA PARK
1907–1913

GREAT WHEEL
2012–Present

PORTLAND

JANTZEN BEACH
1928–1970

LOTUS ISLE
1930–1932

COUNCIL CREST
1907–1929

OAKS AMUSEMENT PARK
1905–Present

SAN FRANCISCO

PLAYLAND-AT-THE-BEACH
1913–1972

| 1900 | 1910 | 1920 | 1930 | 1940 | 1950 | 1960 | 1970 | 1980 | 1990 | 2000 | 2010 | **WASHINGTON** |

White City | Bellingham
Luna Park | Seattle
Natatorium Park | Spokane
Playland | Seattle
Santafair | Federal Way
Lollipop Park | Bellevue
Fun Forest | Seattle
Great Wheel | Seattle

| 1900 | 1910 | 1920 | 1930 | 1940 | 1950 | 1960 | 1970 | 1980 | 1990 | 2000 | 2010 | **OREGON** |

Council Crest | Portland
Jantzen Beach | Portland
Lotus Isle | Portland
Pixieland | Otis Junction
Thrill-Ville USA | Turner
Oaks Amusement Park | Portland

| 1900 | 1910 | 1920 | 1930 | 1940 | 1950 | 1960 | 1970 | 1980 | 1990 | 2000 | 2010 | **NORTHERN CALIFORNIA** |

Idora Park | Oakland
Luna Park | San Jose
Neptune Beach | Alameda
Playland-at-the-Beach | San Francisco
Pacific City & Coyote Point Park | San Mateo
Santa's Village | Scotts Valley
Tahoe Amusement Park | Lake Tahoe
Marine World / Africa USA | Redwood City
Frontier Village San Jose
J's Amusement Park & Haunted House | Guerneville
Boomers | Fresno

● CLOSED ● OPEN

SEATTLE CENTER'S FUN FOREST

Fun Forest was originally built for the 1962 World's Fair as the Gayway Amusement Park. The land developed into Seattle Center's Fun Forest, which remained under the same ownership until it closed in 2011. The little amusement park sat in the middle of the Seattle Center next to the Space Needle and the Experience Music Project (now the Museum of Pop Culture).

Over its nearly fifty-year run, Fun Forest had many attractions and thrill rides, including the Flight to Mars, Wild Mouse, Galaxy, Windstorm, Wild River, and Orbiter. When it closed, the park was down to a dozen rides (about half of which were kiddie rides) and a handful of carnival games.

In the mid-1980s, Disney proposed a multi-million-dollar deal to purchase the park, which would have included a complete update of the Seattle Center. The deal fell through, although later, in 1997, the park was expanded to include the Entertainment Pavilion. The large indoor area included an arcade, laser tag, miniature golf, and a twenty-five-foot climbing wall.

The Fun Forest reportedly closed due to a combination of patron loss, financial strain, and issues with their lease. On January 2, 2011, Fun Forest closed its doors and many longtime Seattleites sadly said goodbye. Today, Chihuly Garden and Glass occupies the space in the Seattle Center. Amusement parks have left Seattle, but you can still ride a 175-foot Ferris wheel, the Seattle Great Wheel, at Pier 57 on Elliott Bay.

COUNCIL CREST, DREAMLAND OF THE NORTHWEST

One of the highest points in Portland, Council Crest offers a view of five Cascade mountain peaks—Mounts Hood, St. Helens, Adams, Jefferson, and Rainier. On what is now a city park, with green space, trails, and vistas, there was once a fully developed amusement park. The Council Crest Amusement Park (a.k.a. Dreamland of the Northwest) opened on Memorial Day in 1907 and closed after Labor Day in 1929.

Similar to other amusement parks in the United States, Council Crest Amusement Park was envisioned by the Portland Railway, Light, and Power Company to attract riders to the streetcar on the weekends and holidays. Most of the park's attractions were built by LaMarcus Thompson from Coney Island and included a canal boat ride, scenic railway, Columbia Gorge riverboat ride, carousel, and Ferris wheel. Other attractions included an observatory, dance hall, Japanese tea pavilion, tavern, and hot air balloon rides. Many postcards of the park have an illustration of the Big Tree Observatory in the center of the park, although there has been no documentation to prove that it was ever built.

The park was not immune to the Great Depression; after a couple of years of losing money, it was forced to close at the end of the season in 1929. The trolley continued to make trips up to Council Crest for twenty years after the park closed. People still visit Council Crest, but for calm mountain views, not mechanical thrills.

PLAYLAND-AT-THE-BEACH

San Francisco's Playland-at-the-Beach was an amusement park on Ocean Beach that operated, in one form or another, from 1896 until August 1972. It started with a collection of rides, entertainers, and shooting galleries, down the hill from the largely popular Sutro Baths and Cliff House, which was built in 1986 by Adolph Sutro. In the early 1920s the park officially became known as Playland-at-the-Beach and a log flume called the Chutes debuted. This was also around the time that the Big Dipper roller coaster opened. It remained a main attraction at the park until 1955, when changes in San Francisco building codes forced the ride to close.

In the late 1920s, George and Leo Whitney took over ownership of the park and remained owners until George's death some forty years later. George Whitney Jr. was recruited by Disney and hired as an Imagineer on the Disney park design team. Whitney Jr. was the only one on the Disney team who had any experience in amusement parks and rides.

For an amusement park that is considered San Francisco's "Lost Treasure," Playland-at-the-Beach's closure was unceremonious. Demolition of the park started just one month after the park closed in August 1972. Today, most of the remnants of the park are those that were looted prior to demolition. The site of Playland now features a different landscape, with apartment complexes and the Great Highway between Cabrillo and Balboa Streets.

JAPAN + FOOD

In these pages, we explore Japanese culinary traditions in San Francisco, Portland, and Seattle. Each city has strong historical ties to Japan. Our research into different types of Japanese cuisine found a deeply woven and complicated story of food and culture. Sushi, ramen, and teriyaki involve different but overlapping stories about the relationship between Japan and the Upper Left.

In the United States, sushi came first, became synonymous with Japanese cuisine, and for many was the gateway food into not just Japanese cuisine, but Japanese culture. Teriyaki, as we know it in the United States, is a hybrid—a mix of Japanese and American culinary traditions—that sprung up in Seattle in the 1970s. The ramen wave of the 2010s is reminiscent of sushi's growth in the 1980s and 1990s. Ramen speaks to the continued and increasing connection that the United States, and the Upper Left, in particular, has with Japan.

The maps on these pages show the distribution of sushi, ramen, and teriyaki restaurants circa 2018. Categorizing restaurants can be tricky, because many restaurants could fall into more than one category. Our goal was to capture the main offerings for each restaurant. We categorized restaurants by using "Best of" lists for each category in each city, and by looking at the names of the restaurants—the logic being that if a restaurant's name included "sushi," "ramen," or "teriyaki," it signaled an emphasis on that particular food. The categorized restaurants on these pages offer only a snapshot in time (as restaurants are notoriously ephemeral) and have been categorized with broad strokes. Even with these limitations, the variation in restaurant types within and between cities is telling.

Sushi restaurants are the most common of the three categories in each city. San Francisco has the most sushi restaurants per capita, followed by Portland, then Seattle. When it comes to ramen, the number of shops per capita follows the same pattern as that of sushi restaurants: San Francisco has the most, Portland closely follows, and Seattle has the least (this time by a larger margin). Seattle has the highest number of teriyaki restaurants per capita. This is no surprise, as Seattle is the birthplace of teriyaki (as we know it in the United States).

HONSHU

SHIKOKU

KYUSHU

JAPAN AND THE
UPPER LEFT, TO SCALE

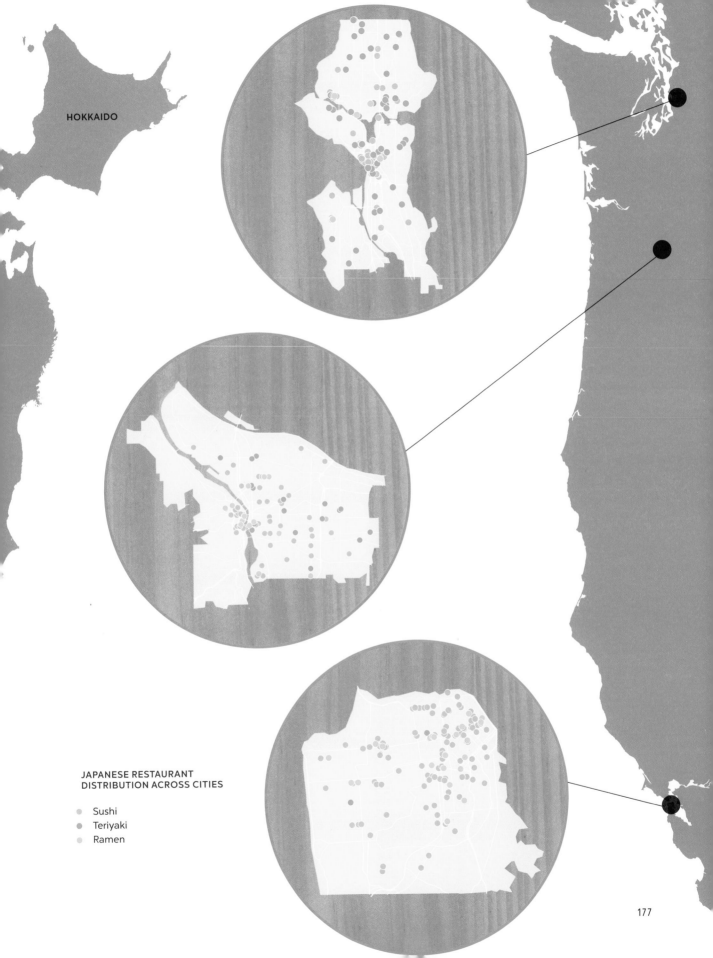

HOKKAIDO

JAPANESE RESTAURANT
DISTRIBUTION ACROSS CITIES

- Sushi
- Teriyaki
- Ramen

177

TERIYAKI

Unlike sushi and ramen shops, teriyaki restaurants did not originate in Japan. US-style teriyaki was created in 1976 by Seattleite Toshihiro Kasahara. Kasahara moved from Ashikaga City, Japan, to Portland in the late 1960s. After graduating from Portland State University with a business degree in 1972, he traveled around the country cooking, eventually opting to settle in Seattle over Portland, deeming it a place with better business opportunities. On March 2, 1976, he opened Toshi's Teriyaki Restaurant on 372 Roy Street in Seattle.

Looking to open a shop that could compete with traditional Japanese restaurants, Kasahara developed a sweet and sticky teriyaki sauce that used sugar instead of the sweet rice wine typically used in Japan. In Japan, teriyaki sauce is a marinade generally used for fish. Toshi's catered to local tastes. His menu consisted of five items, three of which were teriyaki chicken, teriyaki beef, and teriyaki steak, and none of which were fish. White rice and cabbage salad dressed with sesame oil and rice wine vinegar accompanied each order.

Business took off and Kasahara opened another shop in 1980. Many imitators throughout Washington, Oregon, and beyond copied his formula, creating a new style of food—one that borrowed from Asian and Hawaiian cuisine, but was born of 1970s Seattle.

Today, teriyaki restaurants in Seattle are in decline, as the new foodie culture moves away from giant portions and high sugar content. In 2007, Seattle was reported to have more than a hundred teriyaki joints. Our research found fifty-six, a far cry from its heyday, though still more than the twenty-four we found in Portland and the three we found in San Francisco. For many, teriyaki is still simply local food. Toshi Kasahara currently operates Toshi's Teriyaki in Mill Creek, just north of Seattle.

RAMEN

Ramen, made from wheat flour, is simply one noodle style in the Japanese noodle universe, which includes udon (thick noodles made from wheat), sōmen (thin noodles made from wheat), soba (made from buckwheat), and shirataki (made from konjac yam). Ramen first gained notoriety in the United States as an instant cup of soup that sustained college students, urban working-class folks, and artists through lean times.

A wave of high-end ramen shops popped up in Southern California during the early 2010s, then onto San Francisco, Seattle, and, most recently, Portland. Evidence of this is the wait for a bowl of soup at the lauded Afuri, one of Tokyo's shops that established a location in Portland during the mid-2010s.

One key difference between Japanese ramen restaurants and US ramen restaurants is price; ramen in Japan is relatively inexpensive, while ramen in the United States tends to be expensive. Now people in the United States expect that a bowl of ramen may cost upwards of fifteen dollars in a nice restaurant, rather than fifty cents for a Styrofoam cup of noodles at home.

(left) Toshihiro Kasahara, innovator of Seattle-style teriyaki, outside of his Mill Creek shop north of Seattle, 2018.

SUSHI

For many who grew up in the United States during the twentieth century, familiarity with Japanese cuisine was limited to sushi. In Japan, sushi went from being a convenient way to preserve fish in rice with vinegar to a food often reserved for special occasions. In 1966, the first sushi restaurant in the United States opened in Los Angeles. At that time in Japan, sushi meant nigiri—a finely cut piece of uncooked fish on a small mound of rice. In Los Angeles in the 1970s, a chef began using avocado and crab instead of tuna, while inverting a maki roll so the seaweed was on the inside—the birth of the California roll. Many shops served rolls in addition to nigiri and by the 1980s, California rolls became part of the US lexicon of Japanese cuisine.

Sushi increased in popularity in the United States throughout the '80s and '90s, and in many places became a sign of sophistication or gentrification. The California roll migrated back to Japan and has become a small part of Japanese food culture, influenced by its own hybridized cuisine.

IZAKAYA

Also on the rise in Portland, San Francisco, and Seattle are *izakaya*. In Japan, an izakaya is simply a small tavern or pub, an informal place to drink and eat after work. Dishes, small and shared, are ordered continually, rather than all at once. Izakaya have long been places for "salarymen" to have a few drinks after work. Though the clientele in Japan has expanded over time—more women and younger people—it remains a place that people use to transition from work to home.

Upper Left cities took izakaya and made them more formal, more precious, and more expensive—a place where one might even get sushi.

(left) Yataimura Maru, Izakaya, Portland, 2019.

JAPANESE RESTAURANT DISTRIBUTION ACROSS CITIES

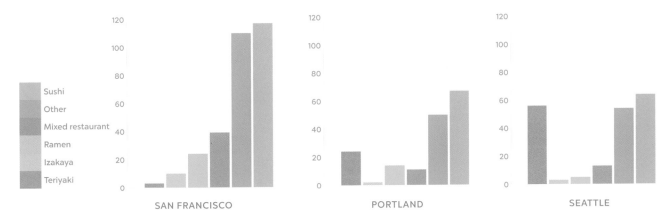

Sushi
Other
Mixed restaurant
Ramen
Izakaya
Teriyaki

SAN FRANCISCO

PORTLAND

SEATTLE

THE ART OF THE POUR:
BEER, WINE, AND SPIRITS

That Upper Left cities offer many opportunities to drink alcoholic beverages is nothing new. Early San Francisco, Portland, and Seattle had plenty of taverns. For many decades, the local tavern was an institution in each city, just as it was in many US cities. Local taverns are not expected to provide any type of alcohol they have made themselves, but rather to provide a steady stream of well-known favorites, with a few exotic options for special occasions. Those taverns still exist, but in some areas are fading in the age of urban breweries, distilleries, and wineries.

We are caught between the era of the dingy pub, which serves Budweiser and comfortable familiarity, and a raft of breweries that all seem to have six or seven varieties of IPA and at least a couple of Belgian sours. And then there is the local distillery that makes the whiskey featuring a portrait of a mutton-chopped man on the label. There is also the urban winery with giant tanks and barrels of wine rising up behind the industrial chic decor and sparsely furnished tasting room. The presence of all these alcohol producers right in the city speaks to the changing urban landscape.

The rise of urban cidery, distillery, and winery scenes are recent phenomena compared to craft brewing. In all three cities, breweries are the most widespread. It takes surprisingly little room to make beer, which is one reason why there are so many breweries. All three cities are close to wine-producing areas. In collecting data on urban wineries, we tried to leave out places that were tasting rooms exclusively, but capture places that produced wine on-site.

SEATTLE

Seattle has the most developed urban winery scene of the three cities. This may have something to do with nearby Woodinville, which hosts dozens of tasting rooms for regional vintners. No need to leave the city anymore. The winery belt stretches along most of the length of the industrial area south of the stadiums. A few more urban wineries are found in West Seattle, and a handful more are scattered in neighborhoods north of downtown. Breweries are plentiful in Seattle and found in just about every part of the city, except the northeasternmost corner (Bitter Lake indeed). Breweries line Salmon Bay, with a tight grouping in Fremont. Another large group of breweries is located in Georgetown. Seattle's developing local spirits scene has a stronghold in the industrial area south of downtown, amid the wineries and breweries. Others are found in Queen Anne, Capitol Hill, and Fremont.

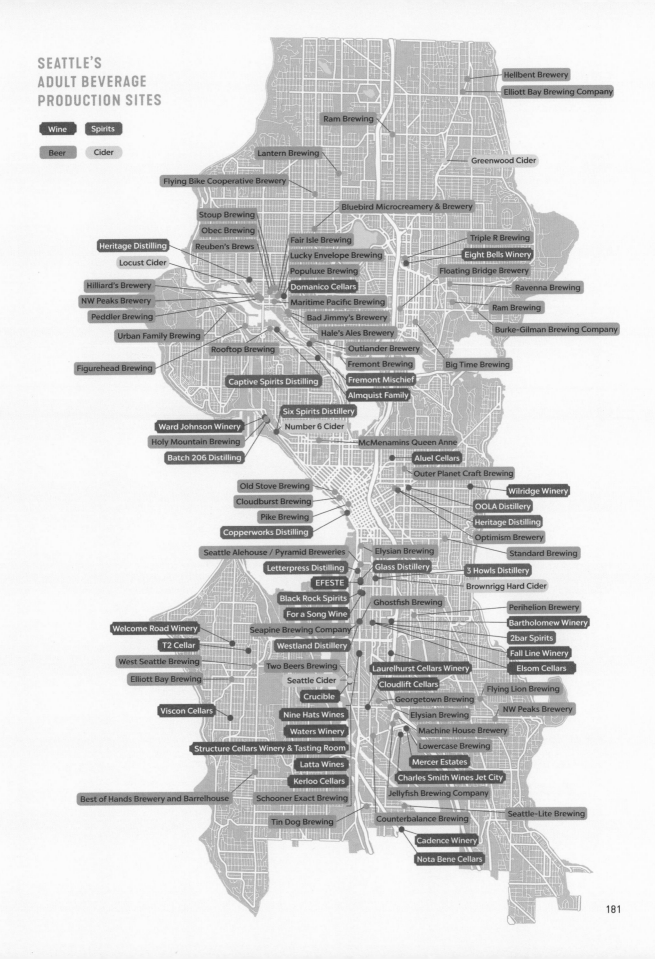

SEATTLE'S ADULT BEVERAGE PRODUCTION SITES

Wine Spirits

Beer Cider

Hellbent Brewery
Elliott Bay Brewing Company
Ram Brewing
Greenwood Cider
Lantern Brewing
Flying Bike Cooperative Brewery
Bluebird Microcreamery & Brewery
Stoup Brewing
Obec Brewing
Fair Isle Brewing
Triple R Brewing
Heritage Distilling
Reuben's Brews
Lucky Envelope Brewing
Eight Bells Winery
Locust Cider
Populuxe Brewing
Floating Bridge Brewery
Hilliard's Brewery
Domanico Cellars
Ravenna Brewing
NW Peaks Brewery
Maritime Pacific Brewing
Ram Brewing
Peddler Brewing
Bad Jimmy's Brewery
Burke-Gilman Brewing Company
Urban Family Brewing
Hale's Ales Brewery
Rooftop Brewing
Outlander Brewery
Figurehead Brewing
Fremont Brewing
Captive Spirits Distilling
Fremont Mischief
Big Time Brewing
Almquist Family
Six Spirits Distillery
Ward Johnson Winery
Number 6 Cider
Holy Mountain Brewing
McMenamins Queen Anne
Batch 206 Distilling
Aluel Cellars
Outer Planet Craft Brewing
Old Stove Brewing
Wilridge Winery
Cloudburst Brewing
OOLA Distillery
Pike Brewing
Heritage Distilling
Copperworks Distilling
Optimism Brewery
Seattle Alehouse / Pyramid Breweries
Elysian Brewing
Standard Brewing
Letterpress Distilling
Glass Distillery
3 Howls Distillery
EFESTE
Brownrigg Hard Cider
Black Rock Spirits
Ghostfish Brewing
Perihelion Brewery
For a Song Wine
Bartholomew Winery
Welcome Road Winery
Seapine Brewing Company
2bar Spirits
T2 Cellar
Westland Distillery
Fall Line Winery
West Seattle Brewing
Laurelhurst Cellars Winery
Elsom Cellars
Elliott Bay Brewing
Two Beers Brewing
Flying Lion Brewing
Viscon Cellars
Seattle Cider
Cloudlift Cellars
NW Peaks Brewery
Crucible
Georgetown Brewing
Nine Hats Wines
Elysian Brewing
Waters Winery
Machine House Brewery
Structure Cellars Winery & Tasting Room
Lowercase Brewing
Latta Wines
Mercer Estates
Kerloo Cellars
Charles Smith Wines Jet City
Best of Hands Brewery and Barrelhouse
Jellyfish Brewing Company
Schooner Exact Brewing
Seattle-Lite Brewing
Tin Dog Brewing
Counterbalance Brewing
Cadence Winery
Nota Bene Cellars

181

PORTLAND'S ADULT BEVERAGE PRODUCTION SITES

- Wine
- Spirits
- Beer
- Cider

Occidental Brewing

Royale Brewing Company

Swift Cider
Breakside Brewery
VanPort Brewing

Funhouse Brews

Look Long Brewery
Jan-Marc Wine Cellars
Mac Wine Cellars
Stormbreaker Brewing
Ecliptic Brewing
Square Mile Cider
Seven Bridges Winery

Old Town Brewing

Great Notion Brewing
Concordia Brewery
Viola Wine Cellars

Level Beer

Widmer Brothers Brewery
Labrewatory
Ex Novo Brewing Co.

Lompoc Brewing

Fire on the Mountain Brewing
Second Profession Brewing

Laurelwood Public House and Brewery
Columbia River Brewing

Boedecker Cellars Winery and Tasting Room
Sasquatch Brewery & New West Cider
Freeland Spirits
Martin Ryan Distilling
Bull Run Distilling
Breakside Brewery

Zanzibar Cellars

Angel Vine Wines

Alter Ego Cider

Baerlic Brewing

Grixsen Brewing

Bow and Arrow Wines
Migration Brewing
Coalition Brewing
Ground Breaker Brewing

Gateway Brewing

Threshold Brewing
Montavilla Brew Works

Brewed by Gnomes
Rosenstadt Brewery
Portland Cider House
Leikam Brewing

Zoiglhaus Portland

Vagabond Brewing
Dirty Pretty Brewing
Stone Barn Brandyworks
Townshend's Distillery

McMenamins Hillsdale
Sasquatch Brewing Company
McMenamins Fulton

Old Market Pub and Brewery

Moonshrimp Brewing

Great Sex Brewing
Ruse Brewing

Division Winemaking Company
Little Beast Brewing
Hopworks Urban Brewery
Teutonic Wine Company
Hip Chicks Do Wine
Gigantic Brewing Company
Portland U-Brew & Unicorn Brewing
13 Virtues Brewing Company

Lucky Labrador Beer Hall

Reverend Nat's Hard Cider Taproom

Upright Brewing Company

Cider Riot!

Fullerton Wines

Cerulean Winery

Culmination Brewing

Rogue Pearl Public House

Natian Brewery

10 Barrel Brewing

Ascendant Beer Company
Pints Brewery

Mt. Tabor Brewing

Back Pedal Brewing

Base Camp Brewing Company

Kells Brew Pub

Wild Roots Vodka

ENSO Winery

Mr Smith's

Vinn Distillery

Cascade Brewing Barrel House

Von Ebert Brewing

Wayfinder Beer
House Spirits Distillery

Coopers Hall

Rogue Eastside Pub & Pilot Brewery

Deschutes Brewery

Modern Times Beer

New Deal

McMenamins Crystal

Hair of the Dog Brewing

Straightaway

Rolling River Spirits

Avid Cider Co.

Eastside Distilling

Lucky Labrador Brew Pub

182

PORTLAND

If Portland has a global reputation for anything, it's beer. The city has more brew festivals than some cities have breweries. These include the Oregon Brewers Festival, the Spring Beer and Wine Fest, the North American Organic Brewers Festival, the Portland International Beerfest, and the Holiday Ale Festival. The artisan spirits scene is concentrated in Southeast Portland, a burgeoning Distillery Row. Cider fans have options too.

SAN FRANCISCO

Breweries are diffuse in San Francisco, but only a handful of breweries operate in the west half of the city. Pockets of urban wineries include South of Market, Dogpatch, and Bayview—all areas undergoing rapid change and gentrification. As of 2019, nine wineries had set up shop on Treasure Island. Small craft distilleries are largely concentrated across the bay in Oakland. Those that operate in the city are in former industrial areas on the east side of town.

SAN FRANCISCO'S ADULT BEVERAGE PRODUCTION SITES

Wine Spirits
Beer Cider

Kendric Vineyards · Oro En Paz
Heartfelt Wines
Tag + Jug Cider Co.
Treasure Island Wines
Treecraft Distillery
Stern Grove Brew
Sottomarino Winery
Winemaker Studios
VIE Winery
The Winery SF
Sol Rouge Winery

Fort Point Beer

San Francisco Brewing
Sufferfest Beer
Woods Polk Station
Holy Craft Brewing
Hammertime
JAQK Cellars

York Creek Vineyards

Bartlett Hall Brewing Co.

ThirstyBear Brewing
21st Amendment Brewery
JAX Vineyards
5 Points
Bluxome Street Winery
Local Brewing Co
Distillery No. 209
Oda Restaurant & Brewery
Anchor Brewing
Hotaling & Co Distillery
Triple Voodoo Brewery
Loos Family Winery
Dogpatch Wineworks
Fallon Place Wines
Harmonic Brewing

Black Hammer Brewing
Pacific Brewing Laboratory
Cellarmaker Brewing
Tank 18

Beach Chalet Brewery
Barrel Head Brewhouse
Magnolia Brewing
Black Sands Brewery

Woods Outbound

Standard Deviant Brewing
Dirty Robot Brew Works

Social Kitchen & Brewery

Woods Cerveceria
A.P. Vin
Southpaw BBQ
The Brew Coop
Almanac Beer
Southern Pacific Brewing Co.
Eristavi Winery
Hop Oast Pub & Brewery

Sunset Reservoir Brewing

Sequoia Sake
Raff Distillerie
Speakeasy Ales and Lagers

Barebottle Brewing
August West Wines
Cellars 33 Winery
Betwixt Wines
Gratta Wines

Seven Stills Brewery & Distillery
Laughing Monk Brewing
Ferment. Drink. Repeat.

8-BIT CITY

In San Francisco, Portland, and Seattle there is a resurgence of mechanical amusement devices, the term used by city governments for arcade games, pinball machines, pool tables, electronic darts, love testers, and other quarter-driven machines, all of which need to be registered. Excluded from this designation are gambling machines, which are regulated differently.

Stepping into an arcade can be like stepping into the past. One of the first coin-operated arcade games appeared in 1972. Pong had two small white bars that acted as paddles that batted a small white square around a black screen. That was it. The comparatively complex Space Invaders was released in 1978, with Asteroids and Galaxian following a year later. The 1980s brought Pac-Man, Ms. Pac-Man, Mario Bros., and Donkey Kong, the classic era of video games.

In the 1970s and 1980s video arcades mattered more. That is no longer the case, as home computers and game systems have graphics that exceed what you can find in an arcade. But now all those people who grew up in the 1970s and 1980s (Gen Xers, you know who you are) are in their forties and fifties and looking to reconnect with the arcade games of their youth. And they are bringing their kids (or grandkids) along for the fun.

Like arcade games, pinball has huge cult followings in Upper Left cities. The first standing electric, mechanical games were developed in the 1930s, when they also became known as pinball machines. Solid-state (or electronic) pinball machines showed up in the mid-1970s. In the late 1990s, video screens were first added, and in the early 2000s fully digital machines were developed.

To make these maps we collected data about the number of mechanical amusement devices from locations around each city and then came up with a density measure for each neighborhood. This provides a basic view of which neighborhoods have the most games.

PORTLAND

Because it is a density measure, a big arcade in a neighborhood with few other gaming places might show up as a lower-level neighborhood. Such is the case with Old Town's Ground Kontrol, which celebrates the golden age of the arcade with over one hundred classic video games and some forty pinball machines. For those on the east side of the Willamette, QuarterWorld offers up over sixty arcade games and thirty pinball machines.

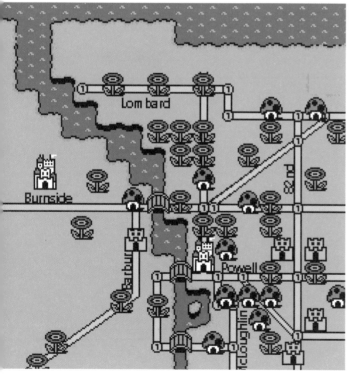

SEATTLE

Seattle also boasts some well-named arcades including Coindexter's in Greenwood, with twenty-six arcade and pinball games, and Flip Flip, Ding Ding! in Georgetown. The Seattle Pinball Museum in the International District has over thirty machines, the oldest of which dates back to 1934. If you like your pinball with ice cream, Seattle chainlet Full Tilt Ice Cream serves up both. And if you like playing arcade and pinball games in an old ice facility, there is The Ice Box in Ballard.

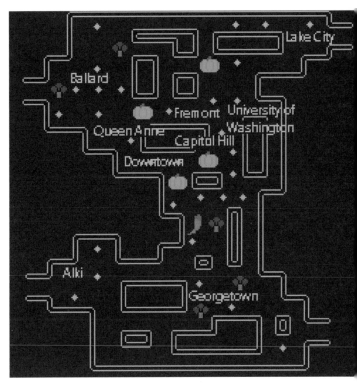

SAN FRANCISCO

Among the many places to play pinball in San Francisco is the fantastically named Free Gold Watch in the Haight, half a block from Golden Gate Park. Free Gold Watch operates out of an old house in a residential neighborhood of Victorians. It is a custom-screen-printing business that also happens to curate the largest collection of pinball machines in San Francisco. Playland Japan in Japantown has a collection of Japanese coin-operated amusement games.

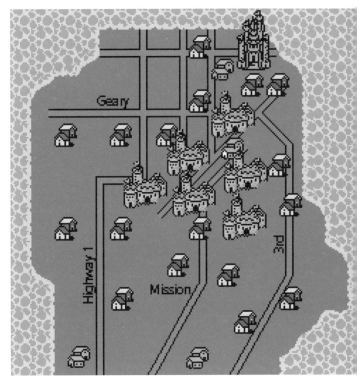

STREET APPEAL: GRAFFITI

Graffiti is considered by some to be a natural form of human expression and an exercise of a collective right to the city. Others consider graffiti a destructive attack upon clean and orderly society.

The context and meanings associated with graffiti change vastly from place to place. As graffiti spreads, it mixes with local traditions of art and writing and becomes something distinct. In the Upper Left, you will find an abundance of very public graffiti on city streets, but also numerous spots in abandoned and secluded spaces.

Each city has its unique graffiti culture, but crews often travel from city to city and spread their graffiti and network with local crews, sharing spots and techniques. Several organizations dedicated to supporting graffiti artists and its culture exist in Upper Left cities, including Endless Canvas (San Francisco), Portland Street Art Alliance (Portland), Alberta Art Works (Portland), and the Graffiti Defense Coalition (Seattle).

Graffiti comes in many forms. On these pages we examine three mediums of graffiti.

STREET ART MEDIUMS

SEATTLE: WHEATPASTES IN POST ALLEY

A curious point of interest is Post Alley, a narrow passage now famous for being both covered in chewing gum and a wheatpaste art wall. Wheatpastes are designs printed or drawn on paper and then applied to walls with liquid glue made from flour or starch and water.

With both gum and wheatpastes, Pike Place Market management and the City of Seattle police take a hands-off approach to these public expressions, allowing and even somewhat encouraging freedom of speech and expression in these spaces.

Even though the city of Seattle's sanitary department cleans off some of the gum bimonthly, they undertook a multiday complete cleaning of both walls in 2015. Within hours of being clean, the gum started to reappear and artists from all over the region descended upon the alley to reclaim the art space with their wheatpastes.

(top) Wheatpaste on Post Alley, Seattle, 2018.

SAN FRANCISCO: STENCILS

Stencil graffiti often seems less shocking and less menacing than many other forms of graffiti. This may be in part because stencils blend seamlessly into the landscape. After all, cities are full of legal stencils, most of which seem to be telling people what not to do—NO PARKING, NO SMOKING, NO LOITERING, NO GRAFFITI.

As many stencils are painted on sidewalks, they are often overlooked and eventually wear away. When done by hand, stencil templates are delicate and complicated to create. However, once created, most stencils can be applied to the landscape quickly and repeatedly. For most people, stencil images tend to be easier to recognize and read than a tag.

San Francisco's Mission District has a rich and long-standing tradition of stencils. We have documented stencils in this neighborhood from time to time since 2006. Many of the stencils are explicitly political and target gentrification and the tech industry. Other stencils carry positive or humorous affirmations.

PORTLAND: STICKERS

Portland has a thriving and artistic graffiti sticker art scene. Some think Portland's wet weather and mural regulations have encouraged artists to focus more on vinyl mediums; others look back to local DIY and skateboarding roots. In other cities you might just see HELLO, MY NAME IS and US postal stickers, but in Portland the stickers stand out.

For more than ten years, the Portland sticker art scene has blossomed into a colorful and relentless force on city streets. To the dismay of tireless graffiti abatement crews, volunteer community groups, and vigilantes, the stickers just keep getting put up. It is an endless cycle of slap, stick, and scrape. The sticker art community in Portland has built such a strong community that they have at times come together to form "sticker crews," such as Visual Assault Crew and Sticker Nerds. Portland even hosts an annual Sticker Nerds Art Show, a temporary art gallery transformed into an alternate reality where every wall and fixture is covered with stickers.

(top) Stencils in the Mission District, San Francisco, 2006–2008 and 2017–2018.

(bottom) Stickers on Hawthorne Street, Portland, 2018.

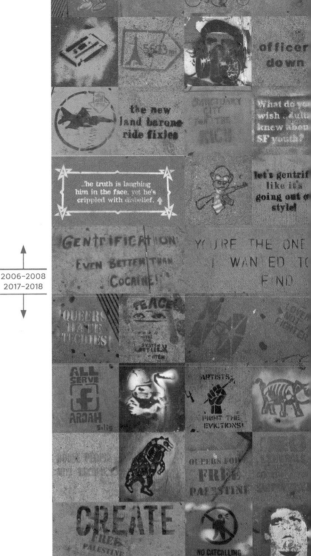

2006–2008
2017–2018

FROM RUINS TO RUIN: FORMER GRAFFITI SPOTS

SEATTLE: TUBS

For years, an old building known as TUBS sat vacant at the corner of Fiftieth and Roosevelt in the University District amid a bustling urban neighborhood. In 2009, the building's owner, thinking the building would soon be torn down, invited graffiti artists to use the twelve-thousand-square-foot space as a canvas for their art. The owner wanted to provide the community with an "ephemeral and evolving" piece of street art. The space became a free wall and a hot spot for Seattle graffiti.

A year after the free wall began, the city had received more than nine hundred graffiti complaints about the property. But the building owner fought back, citing private property rights and community appreciation for the art. By this point, TUBS had become a tourist destination, and like many graffiti meccas, served as an urban backdrop for photographers and filmmakers.

The city of Seattle claimed it had no authority to force the owner to clean up the building. Since the city of Seattle defines graffiti as "unauthorized markings" and the TUBS owner willingly allowed the building to be marked, the city of Seattle could not issue a fine. The free wall at TUBS continued

until 2014, when it was finally demolished to make way for a large condo building. The TUBS free wall was an important piece of Seattle's urban art history.

PORTLAND: PIRATE TOWN

While its official name is Triangle Park, it has been known by many names: Popsicle Land, Creosote Factory, SuperFun Site, and most infamously, Pirate Town. Located on the Portland Harbor at the base of Waud Bluff in the University Park area of North Portland, this abandoned complex provided both a canvas for graffiti writers and a space for the creation of DIY skateboarding and BMX structures.

Pirate Town was the site of the former Riedel North Portland Yard, which dredged rivers, constructed boats, and cleaned up hazardous railroad spills. Riedel closed in 1986, but the effects of its operations (and the site's prior operations) will be present for centuries to come; soil and groundwater tests show high levels of toxic contaminants, mainly arsenic. It is now a Superfund site.

Skaters and BMX riders revamped Pirate Town. The complex was covered in layers of paint. For years, Pirate Town was a cherished space for all sorts of adventure; it was a horror movie theater, an army training ground, and a place to end

(top, left and right) TUBS, Seattle, 2013.

midnight bicycle rides, hold massive parties, and host epic chariot wars.

In December of 2008, the University of Portland bought the site for $6 million and swiftly demolished it, releasing a statement saying that the space was a liability. As of this writing, the site remains vacant, largely reclaimed by nature.

SAN FRANCISCO: COSSON HALL, TREASURE ISLAND

Treasure Island is an artificial island in the bay about halfway between Oakland and San Francisco. It is connected by a causeway to Yerba Buena Island, which connects the east and west parts of the Bay Bridge. Treasure Island was constructed to host San Francisco's third world fair—the 1939 Golden Gate International Exposition. The Navy opened a station there in 1941.

Built in 1969, Cosson Hall consisted of two six-winged naval barracks, occupied by sailors stationed on Treasure Island. From above the structures looked like giant asterisks. Abandoned in the early 1990s, the buildings fell into extreme disrepair and were covered with extraordinary layers of colorful graffiti. An open-air interior building design allowed writers to paint in broad daylight and go unseen.

Rumors suggested the barracks were haunted. Demolished in 2016, the buildings themselves are now ghosts.

(above, top) Pirate Town, Portland, date unknown.

(above, middle) Pirate Town, Portland, 2005.

(below, left and right) Cosson Hall, Treasure Island, San Francisco, 2014.

THE BUFF: GRAFFITI ABATEMENT

Every year millions of dollars are spent on graffiti abatement—washing off, covering up, or otherwise getting rid of graffiti. What makes graffiti distinct from other script or images in the city is its legal status. Art done on a property without the consent of that property's owners is considered vandalism. With consent, the same work might be called a mural, public art, or advertising. Because commercial interests do not control graffiti, the content is often different and distinct from sanctioned messages on urban landscapes.

The broken-windows theory, first introduced in a 1982 *Atlantic* article, has heavily influenced graffiti abatement across the United States, including in the Upper Left. The article argued that things such as broken windows, left unrepaired, broadcast to

(top) Buff in Portland, 2019; (bottom) Buff in San Francisco, 2008.

the public that an area is uncared for and uncontrolled, which will eventually catalyze worse and possibly violent crimes. Scant academic research supports the "broken windows" claim of a direct link between petty crimes and violent crimes.

Increasingly, cities in the Upper Left are asking their residents to report graffiti to help eliminate it from the landscape—a policy of zero tolerance. In the first decade of the 2000s, cities set up hotlines to report graffiti. Smartphone apps, developed in the 2010s, make it easier to report graffiti. These reporting mechanisms allow cities to gather more data, which we've used to make the maps on the next three pages.

These are maps of where people reported graffiti (2017) in San Francisco, Portland, and Seattle—*not* maps of all the graffiti in Upper Left cities. Some areas that have a lot of graffiti—industrial areas, for example—do not show up much on these maps, because so few complaints are made about graffiti in those locations. Likewise, an area with a dense concentration of complaints may not necessarily be an area with a lot of graffiti—a single piece of graffiti may have generated most of the complaints.

Cities spend a lot of money abating graffiti. Tracking those expenditures is difficult; yearly statistics are had to come by. Our research shows that in 2016, Portland spent approximately $1.8 million abating graffiti, which is triple the figure from 2011. The most recent statistics available for Seattle indicate about $2.5 million was spent abating graffiti in 2009.

In 2014, the San Francisco Budget and Legislative Analyst Office reported that the city's total abatement costs for the previous fiscal year was more than $24 million. These figures correspond to covering up and removing graffiti and do not include the legal costs of arresting or prosecuting graffiti writers.

DENSITY OF REPORTED GRAFFITI COMPLAINTS

Low High

DENSITY OF REPORTED GRAFFITI COMPLAINTS

Low   High

DENSITY OF REPORTED GRAFFITI COMPLAINTS

Low High

GOING THE DISTANCE: MARATHONS

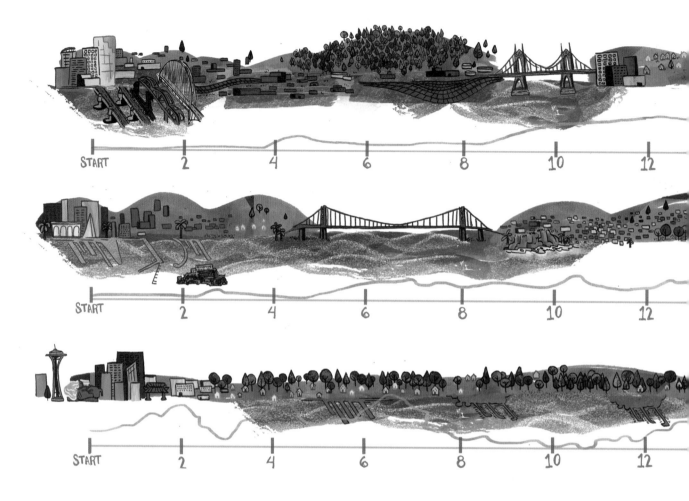

Marathons are based on the story of Greek soldier Pheidippides (or Philippides), who in 490 BCE ran nearly twenty-five miles from Marathon to Athens to deliver news of victory over Persia and then promptly collapsed and died. To commemorate this event and add a sense of drama to the proceedings, the modern Olympics included the marathon—which was never part of the ancient games—as the final event. The original race measured just shy of twenty-five miles and only nine of twenty-five participants finished the grueling event.

The length of the marathon was not standardized until the London Olympics of 1908, when a twenty-six-mile course was lengthened another 385 yards so that runners could finish the race by running past the queen in her royal box. To this day, official marathons maintain this twenty-six-mile-plus standard, regardless of royalty in attendance.

Established in 1897, the elite Boston Marathon is the oldest continuous marathon in the world. Prospective runners must meet qualifying times in other marathons before being allowed to enter a lottery to participate. The Portland, San Francisco, and Seattle Marathons are all Boston Marathon qualifying races.

By the mid-1990s, cities across the world had established marathons. These races feature

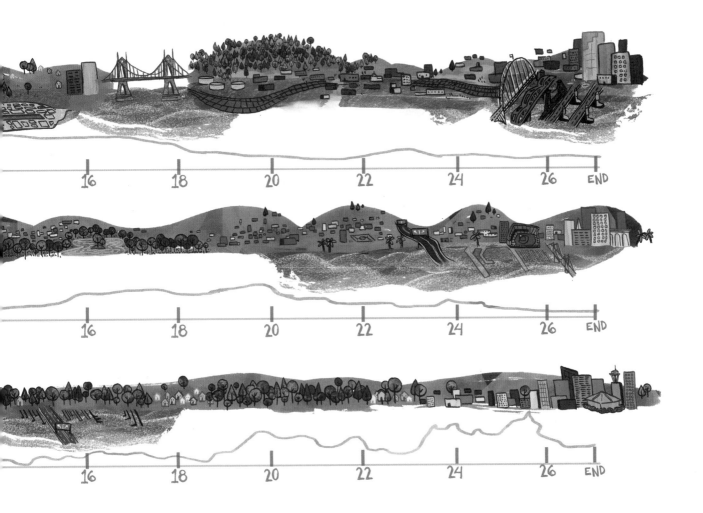

and celebrate the city. Increasingly, courses are designed to engage runners and viewers in a visual journey. Each race is a tour of the city and draws tourist spectators.

Fitting the twenty-six-mile stretch into San Francisco requires making a loop. The loop runs along the coastline for about half the race, across the Golden Gate Bridge and back, and finally through Golden Gate Park and historic neighborhoods. The trick in San Francisco is to keep the runners from going up too many hills.

The Seattle Marathon is also a loop, but with a tail. The loop starts near the Space Needle, takes runners through downtown amid the towering buildings, out to Lake Washington, south through Seward Park, then back north, ending in Memorial Stadium.

The Portland Marathon was a loop in 2016, but a snafu on the day of the race resulted in a number of participants veering off course for an additional half mile. This led to a change in the course for 2017. The new route is an out-and-back course, starting at Tom McCall Waterfront Park, heading north through the industrial district, crossing St. Johns Bridge, and winding through Overlook neighborhood before turning around.

STADIUMS, PALACES, AND DOMES

From the early years, cities have had spaces for sport, whether it be an open field, a dirty alley, or a bumpy street. Early playing fields (and later the earliest stadiums) were open, informal, and located close to where people lived, so it was an easy walk from home to field. Gradually, workers enjoyed more leisure time and disposable income, which fueled the rise of professional sports, notably baseball. Teams moved so frequently that stadiums were often hastily constructed and built from wood, making them susceptible to fire.

Enclosing baseball parks was also part of the commercialization of the game—admission could be charged and those who didn't pay were kept out (mostly). Over time, usage of *park*, *field*, and *grounds* faded in favor of the term *stadium* or, more dramatic still, *coliseum*. From the early fields to the steel and concrete venues of the 1920s to the cookie-cutter stadiums of the 1960s and 1970s, then to the return of single-sport venues in the urban core, we take a closer look at sport venues.

NINETEENTH-CENTURY VENUES

Up until the late 1880s, blood sports were a major form of entertainment in San Francisco. Modern sports began to displace more traditional diversions, as the elite considered baseball more respectable than cockfights, bull versus bear matches, and prizefighting.

The first pickup baseball game in the Upper Left reportedly took place in San Francisco's Portsmouth Square in 1851. However, the first *official* game took place in the Mission, at Sixteenth and Harrison, in 1859. The Mission became the site of several venues, including the Recreation Grounds, which opened in 1868 to become the first enclosed baseball park on the West Coast. Today it is Garfield Square. By the 1870s, undeveloped areas in the Mission and South of Market served as amateur and practice fields—spaces that became known as sandlots.

SEATING CAPACITY AND STADIUM TYPE

○ <5,000
○ 5,000+
○ 10,000+
○ 20,000+
○ 40,000+
○ 60,000+

— Open
— Closed
---- Former size

● Former sport
● Current sport

⚾ Baseball   🏀 Basketball
🥊 Boxing   🏈 Football
🏒 Hockey   🏁 Racing
⚽ Soccer

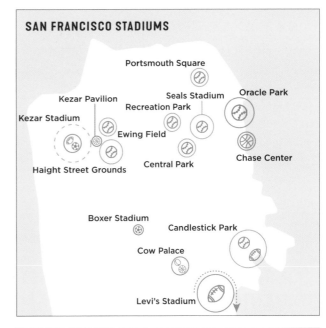

SAN FRANCISCO STADIUMS

Portsmouth Square
Kezar Pavilion
Kezar Stadium
Seals Stadium
Recreation Park
Oracle Park
Ewing Field
Chase Center
Haight Street Grounds
Central Park
Boxer Stadium
Candlestick Park
Cow Palace
Levi's Stadium

PORTLAND STADIUMS

Rose City Speedway
Portland International Raceway
Woodburn Dragstrip
Merlo Field
Exposition Rink
Veterans Memorial Coliseum
Vaughn Street Park
Rose Garden/Moda Center
Providence Park
First Regiment Armory Annex
Charles B. Walker Stadium
Ron Tonkin Field
Keller Auditorium
Portland Ice Arena

SEATTLE STADIUMS

Husky Stadium
Climate Pledge Arena
Civic Field
Lumen Field
Kingdome
Sicks' Stadium
T-Mobile Park
Dugdale Field
Starfire Sports Complex
ShoWare Center

In 1884, Central Park opened on the corner of Eighth and Market Street, later becoming home to the San Francisco Seals. The Haight Street Grounds near Golden Gate Park opened in 1886 and could hold twenty thousand spectators. Recreation Park, built in the late eighteen hundreds on Valencia Street, hosted the city's first Pacific Coast League game. All three stadiums were destroyed by the 1906 earthquake and fire.

The beginning of amateur baseball in Portland started in 1866 with the Portland Pioneers, a company team. Multnomah Field was one of the earliest outdoor sites in Portland dedicated to sport. Built in 1893 by the Multnomah Amateur Athletic Club (MAAC), the field's grandstands and clubhouse were destroyed by fire in 1909 and later rebuilt.

Seattle's first professional baseball game was reportedly held on May 24, 1890, at Madison Park, when the Seattles defeated a team from Spokane. Also home to the Hustlers, Madison Park hosted baseball games until 1892. A key figure in early Seattle baseball was journeyman big-league player Dan Dugdale, who moved to Seattle in 1898, lured by the gold rush. He never left for the Klondike, deciding to pursue baseball as a business by buying the Seattles and renaming them the Braves. Dugdale went on to own the Klondikers, the Clamdiggers, the Chinooks, the Turks, and the Giants.

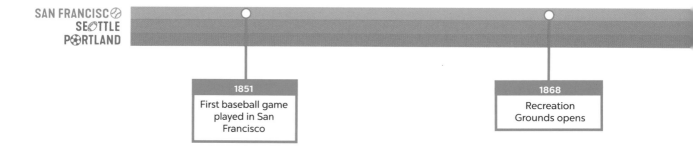

1851
First baseball game played in San Francisco

1868
Recreation Grounds opens

EARLY TWENTIETH-CENTURY VENUES

Efforts to rebuild San Francisco after the 1906 earthquake included creating new spaces for baseball. In 1907, a fifteen-thousand-seat Recreation Park opened and hosted the Seals through 1913. In 1914, the Seals moved to Ewing Field, a new eighteen-thousand-seat home in the Richmond District. Players and fans found the location miserably windy, cold, and foggy, so the Seals and their supporters returned to Recreation Park the following season. Ewing Field was, however, an important venue for amateur baseball and one of the few places amateur and semipro Japanese baseball teams could play. The stadium burned down in June 1926. On the site currently sits a residential development, Ewing Terrace.

Portland's first open-air stadium, Vaughn Street Park, opened in 1901 on the edge of the emerging industrial area in the northwest part of town. Two rival streetcar line operators constructed the venue in order to boost ridership. The park featured the Giants, who became the Beavers in 1906. Most professional leagues at this time fielded only white players. An attempt to break the color barrier came in 1914, when a Beavers' manager tried to sign Lang Akana, a player of Hawaiian and Chinese heritage. Beavers players threatened to quit if Akana signed, so the signing never happened.

The color barrier in Portland professional baseball was finally broken in 1949, when Frankie Austin and Luis Marquez became the first Latino players to sign with the Beavers. In 1946, Vaughn Street Park was home to the Rosebuds, the West Coast Negro Baseball League team owned by Olympic gold medalist Jesse Owens. By the early 1950s, fires and vandalism had left Vaughn Street Park in dire shape.

Back in Seattle, Dugdale opened Yesler Way Park on Twelfth and Yesler, a six-thousand-seat venue demolished just six years after its construction in 1907. His next venue, Dugdale Field, opened in Rainier Valley in 1913. The eight-thousand-seat venue was the first double-decker baseball venue on the West Coast. A Fourth of July fire destroyed Dugdale Field in 1932.

Hockey venues in Portland and Seattle opened during this era as well. In 1914, the Portland Ice Arena (a.k.a. the Portland Hippodrome) opened on NW Marshall, hosting hockey teams the Rosebuds, the Buckaroos, and the Eagles (who for a year were the Penguins).

The Seattle Ice Arena, on University Street and Fifth Avenue, opened in 1915. With a capacity of four thousand, the venue hosted the Seattle Metropolitans, who in 1917 became the first American team to win the Stanley Cup (take that, 1928 New York Rangers!). The 1917 and 1919 Stanley Cup Finals were held in the Seattle Ice Arena.

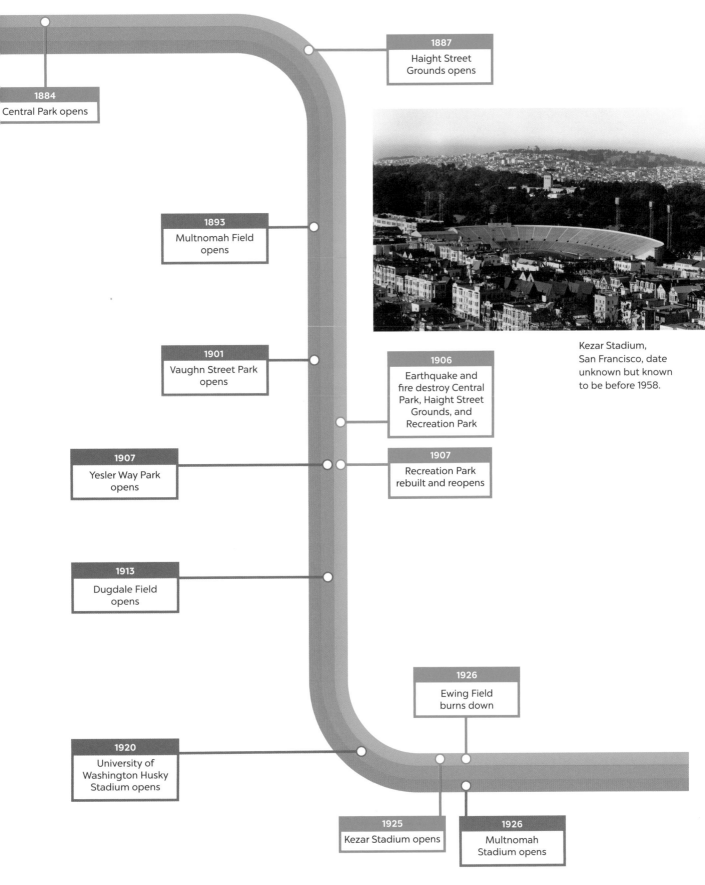

1884
Central Park opens

1887
Haight Street Grounds opens

1893
Multnomah Field opens

Kezar Stadium, San Francisco, date unknown but known to be before 1958.

1901
Vaughn Street Park opens

1906
Earthquake and fire destroy Central Park, Haight Street Grounds, and Recreation Park

1907
Yesler Way Park opens

1907
Recreation Park rebuilt and reopens

1913
Dugdale Field opens

1926
Ewing Field burns down

1920
University of Washington Husky Stadium opens

1925
Kezar Stadium opens

1926
Multnomah Stadium opens

LARGER, MORE MODERN STRUCTURES

In San Francisco, Kezar Stadium opened in 1925 at the east end of Golden Gate Park. The sixty-thousand-seat venue was home to many teams from many different sports, most famously the 49ers from 1946 to 1970. Even the Oakland Raiders played at Kezar for part of the 1960 season, while the Coliseum was being built across the bay. In 1989, the original Kezar Stadium was demolished and rebuilt as a ten-thousand-seat venue.

In 1931, Seals Stadium opened in the Mission. Bounded by Bryant Street and Sixteenth Street, the venue was home to the Mission Reds through 1937, the Seals through 1957, and the San Francisco Giants in 1958 when they first moved from New York. Today, the site is a shopping center and parking lot. Still in operation is the nearby Double Play Bar and Grill, which used to serve thirsty fans (and apparently a few players) before, during, and after games.

Although technically located in Daly City just across the San Francisco boundary, the Cow Palace is an important part of San Francisco's sport history. Opened in 1941 as the California State Livestock Pavilion, the Cow Palace served as the site of the 1956 and 1964 Republican National Conventions, in addition to hosting many other boxing matches, rodeos, and circuses. Still in operation, the Cow Palace served as home to the NBA's San Francisco Warriors for most of 1962 to 1971, and also hosted hockey, indoor soccer, and indoor football.

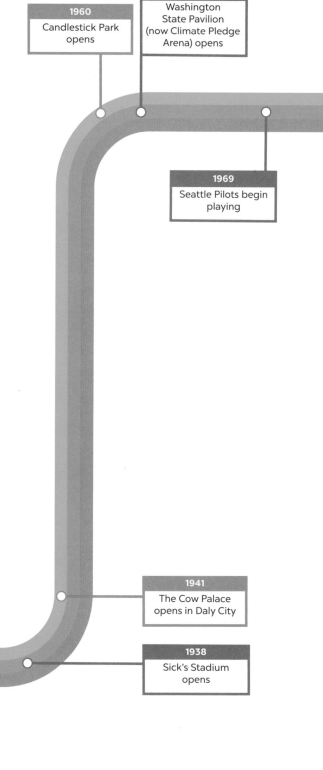

1960
Candlestick Park opens

1962
Washington State Pavilion (now Climate Pledge Arena) opens

1969
Seattle Pilots begin playing

1931
Seals Stadium opens

1941
The Cow Palace opens in Daly City

1938
Sick's Stadium opens

1932
Dugdale Field burns down

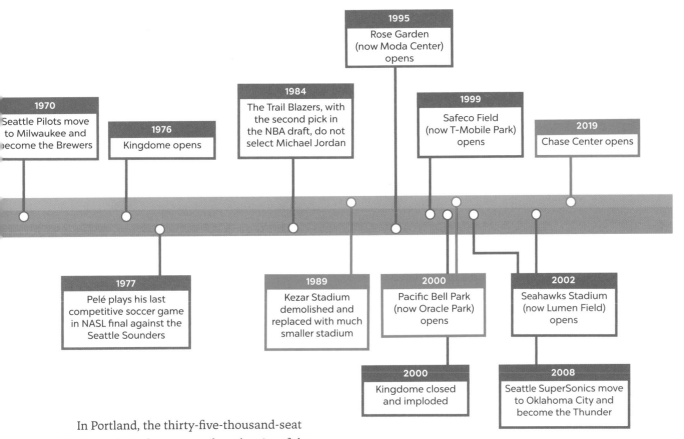

1970
Seattle Pilots move to Milwaukee and become the Brewers

1976
Kingdome opens

1984
The Trail Blazers, with the second pick in the NBA draft, do not select Michael Jordan

1995
Rose Garden (now Moda Center) opens

1999
Safeco Field (now T-Mobile Park) opens

2019
Chase Center opens

1977
Pelé plays his last competitive soccer game in NASL final against the Seattle Sounders

1989
Kezar Stadium demolished and replaced with much smaller stadium

2000
Pacific Bell Park (now Oracle Park) opens

2002
Seahawks Stadium (now Lumen Field) opens

2000
Kingdome closed and imploded

2008
Seattle SuperSonics move to Oklahoma City and become the Thunder

In Portland, the thirty-five-thousand-seat Multnomah Stadium opened on the site of the Multnomah Field in 1926. From 1933 to 1956, the venue featured a dog-racing track, which was removed for baseball, sending the greyhounds to Portland Meadows.

Seattle's Sick's Stadium opened in 1938 on the Dugdale Field site in Rainier Valley. Financier Emil G. Sick bought the Indians, renamed them the Rainiers after his brewery, and paid for the 14,600-seat stadium's construction. The Steelheads of the West Coast Negro Baseball League of 1946 also played home games at Sick's Stadium.

After Emil's death in 1964, his family changed the stadium name from Sick's to Sicks' Stadium and sold it to the city in 1965. In 1969, Seattle was awarded the Seattle Pilots, an MLB expansion team, with the stipulation that Sicks' Stadium serve only as a temporary home for the team. However, after one season, the Pilots fell into bankruptcy and moved to Milwaukee to become the Brewers.

(right) The implosion of the Kingdome, Seattle, March 26, 2000.

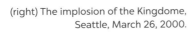

STADIUMS OF THE MODERN ERA

Construction of stadiums slowed after World War II as value of land increased with the rising post-war economy. Stadiums that opened at this time tended to be on the urban fringe. Such is the case with Candlestick Park, home to the San Francisco Giants from 1960 to 1999 and the San Francisco 49ers from 1971 to 2013. The damp, cold, and windy point on San Francisco Bay was chosen as the site of the stadium in part because there was room for ten thousand parking spaces. The Oakland Raiders played some of the 1960 season and all of the 1961 season there. The Beatles played their final concert at Candlestick Park in 1966. The venue closed in 2014 with a capacity of nearly seventy thousand and was demolished the following year.

In Portland, the Lloyd District's Veterans Memorial Coliseum opened in 1960. This was the first home of the Portland Trail Blazers. Since 1976, the venue has hosted the Winterhawks—a junior ice hockey team. In 2009, the glass and concrete structure was added to the National Register of Historic Places.

The city of Portland purchased Multnomah Stadium in 1967 and renamed it Civic Stadium. In 1973, after seventy years in Portland, the Beavers relocated to Spokane. That year, TV actor Bing Russell and his son Kurt (not yet famous) established the Portland Mavericks of the Single-A Northwest League. A rowdy reputation and a cast of quirky players made the team a spectacle until it folded in 1977. The Timbers played in the venue from 1975 to 1981 as part of the North American Soccer League.

Seattle's KeyArena started life in 1962 as the Washington State Pavilion then became the Washington State Coliseum, before being renamed the Seattle Center Coliseum. KeyBank bought the naming rights in 1995, the year of a major renovation. The venue was home to the SuperSonics of the NBA until 2008, when the franchise moved to Oklahoma City to become the Thunder. The Seattle Storm of the WNBA called KeyArena home from 2000 to 2018, when the arena closed for yet another renovation. In 2020, KeyArena was renamed Climate Pledge Arena and will be home to the Seattle Kraken NHL team.

Holding fifty-nine-thousand spectators and featuring an eleven-acre concrete roof, the Kingdome opened in 1976. The venue was home to the Mariners, the Seahawks, the Sounders, and the SuperSonics. In 2000, after standing for twenty-four years, the Kingdome was imploded to make way for a new football stadium. Guinness World Records recognized the Kingdome as the largest building (by volume) ever imploded. Fifteen years after its demolition, King County taxpayers finished paying debt on the venue.

STADIUMS FOR THE TWENTY-FIRST CENTURY

San Francisco's newest large sport venues were built in the industrial areas by the bay. In 2000, Pacific Bell Park, the forty-one-thousand-seat new home of the San Francisco Giants, opened on its site in China Basin. It was renamed AT&T Park in 2016, and then Oracle Park in 2019. The new eighteen-thousand-seat home of the Golden State Warriors, the Chase Center, opened nearby in Mission Bay.

In Portland, the site that started out as Multnomah Field in 1893 and then Multnomah Stadium in 1926 continues to be expanded and renamed. The Timbers returned to the stadium in 2001. The venue, now called Providence Park, also hosts women's soccer team the Thorns. Construction of a second deck on the west side of the stadium, expanding capacity another four thousand seats, finished in 2019. Across the street is the Cheerful Bullpen, a baseball-era bar dating back to at least the early 1970s, and unlike the stadium, still has pitchers.

SAN FRANCISCO

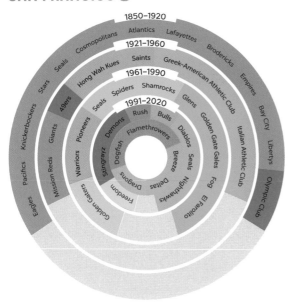

In 1995, the Rose Garden opened, becoming the home of the Portland Trail Blazers. In 2013, a health insurance company purchased the naming rights to the 19,400-seat venue, renaming it the Moda Center.

In Seattle, the new home of the Mariners, Safeco Field (renamed T-Mobile Park in 2018), opened in 1999 south of downtown next to the Kingdome. The venue cost $517.6 million to construct, more than 70 percent of which came from public funds. The venue features a retractable roof, which covers, but does not seal, the venue. In 2002, Seahawks Stadium (renamed Qwest Field in 2004) opened right next door to Safeco and the Seattle Seahawks returned from their temporary home in Husky Stadium. Currently called Lumen Field, the venue has also been home to the Seattle Sounders since 2003.

SEATTLE

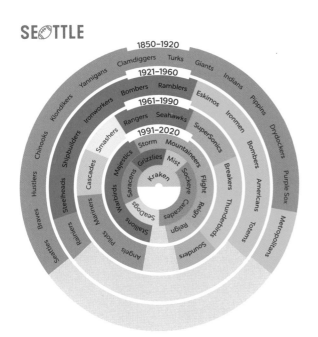

PORTLAND

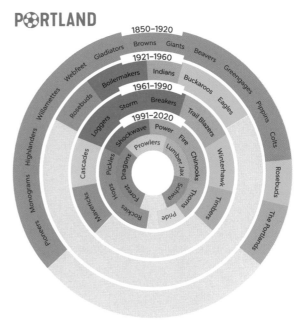

HIT THE BALL!	TO THE END ZONE!	INTO THE NET!
BASEBALL	FOOTBALL	BASKETBALL
TENNIS	ARENA FOOTBALL	HOCKEY
CRICKET	INDOOR FOOTBALL	SOCCER
VOLLEYBALL	ULTIMATE FRISBEE	INDOOR SOCCER
	AUSTRALIAN FOOTBALL	LACROSSE
	RUGBY	

REVERT TO TYPE

Most of the maps in this book were made with computers and only with computers. There are a few exceptions. Three of them are on this page.

These maps were made with a Hermes Rocket 3000 typewriter (made in Brazil) and have the mistakes and unevenness of typewritten work—when done by someone who hasn't used a typewriter for a long time.

We cut out the shape of a city, took the page with the city-shaped hole in it, and placed another piece of paper behind it. The two pieces were slid

into the typewriter and the names of each city were typed repeatedly over the cut-out portion. When the top sheet was removed, the typed impressions of each city remained.

Even here, the simplified city outlines were originally generated by a computer. The detailed invisible lines of the outer digital border stand in contrast to the raw serif typeface repeated inside. There is a texture to analog things that is difficult to duplicate digitally.

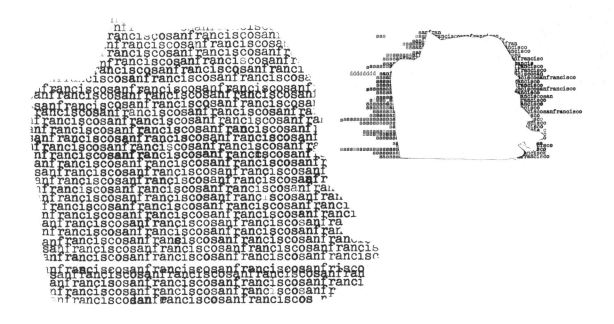

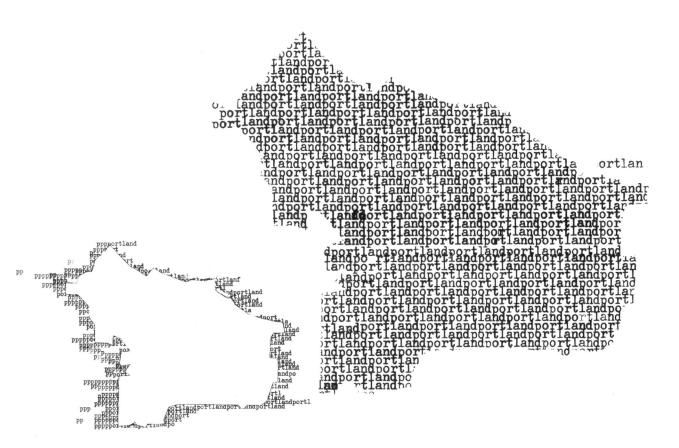

NOTED PLACES

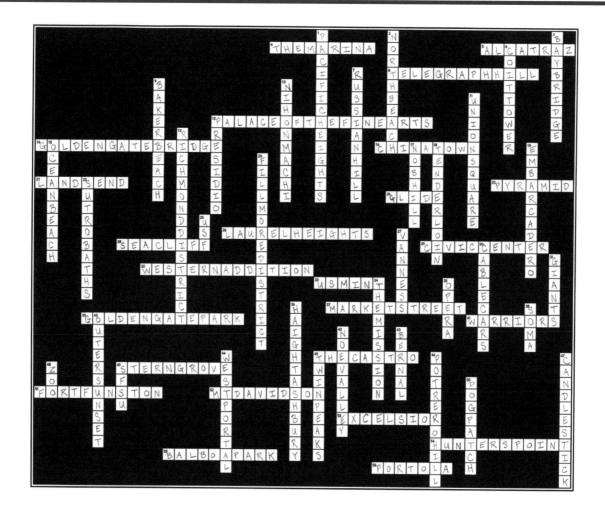

ACROSS

4 Home to "Singles Safeway"
5 Capone slept here
8 Home to feral parrot colony
12 Only surviving 1915 expo structure
14 Its official color is international orange
16 Most densely populated area west of Manhattan
21 Offers Golden Gate Bridge views and a hidden labyrinth
23 City's tallest building until 2018
24 LGBTQ-affirming Methodist church
26 Once a cemetery, now a medical school
28 Robin Williams lived here
29 Home to politics, art, opera, books, symphony, body snatchers
32 Formerly platted as five hundred square blocks
33 Old one closed in 1937; this is the new one
37 Site of city's first horse-drawn rail lines
39 Place of drum circles, windmills, buffalo, and polo fields
41 Old home, Cow Palace. New home, Chase Center
45 Location of rainbow crosswalks
49 Golden Gate Park amphitheater
51 Parking lot, former missile launch site
52 Topped by Armenian memorial
53 Used to be called China Street
54 Decommissioned naval shipyard
55 Ocean runs through it
56 Home of Maltese consulate

DOWN

1 Nancy Pelosi's home
2 City Lights bookstore
3 Half named for Willie Brown
6 Built on the site of former telegraph station
7 Kirk's home in *Star Trek II & III*
9 First site of Burning Man
10 Cherry Blossom Festival held here
11 Site of opening scene in *The Birds*
12 Former Army base, current National Recreation Area
13 Has an "inner" and an "outer"
15 Former home to Playland
17 Grace Cathedral and luxury hotels
18 Named for gritty NYC neighborhood
19 Built on landfill . . . but no more freeway!
20 Former hub of West Coast jazz
22 Seven public swimming pools now in ruins
25 Lone Mountain school
27 Named after San Francisco's seventh mayor
30 A way to get around since 1873
31 Arrived from NY in 1958
34 Home to oldest building in the city
35 SF Sopranos
36 Summer of Love HQ
38 Moscone Center digs
40 Hope you like fog
41 Home of the Gators
42 Named for last Mexican mayor of Yerba Buena
43 Home to state's oldest farmers market
44 Named for streetcar station
45 Fog blockers
46 Dirty Harry's home
47 Built for baseball, closed hosting football
48 Roar!
50 From working-class to biotech

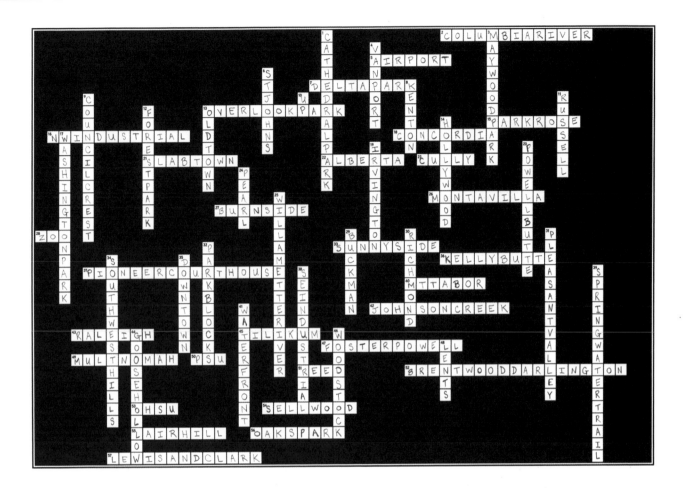

ACROSS

2. Roll on, Woody Guthrie
5. Famous carpet
7. Home of Portland International Raceway
13. Sweet Willamette River views
15. Site of Rossi Farms
16. Smallest number of residents of any neighborhood in Portland
18. Neighborhood named after university
21. Former site of lumber mill, hippodrome, and stadium
22. Widely known for gentrification
23. Site of Chinook village called Neerchokikoo
26. Short for Mount Tabor Village
27. Divides city north and south
28. Deepest transit station in North America
33. Home to Peacock Lane Christmas lights
36. Former site of municipal Isolation Hospital
37. Second-oldest federal building west of the Mississippi
40. Site of rad adult soapbox derby in August
42. Twenty-five-mile tributary of the Willamette
43. Location of first New Seasons Market
45. First major bridge in the United States not designed for cars
47. A.k.a. FoPo
49. The Village
50. Where this book was made
51. Dr. Demento's alma mater
52. Home to Learning Gardens Laboratory
53. Campuses connected by aerial tram
54. Features an amusement park next to a wildlife refuge
55. Former Jewish and Italian neighborhoods
56. Roller-skating year-round since 1905
57. Coed from its 1870s beginning

DOWN

1. Named for Gothic arches under St. Johns Bridge
3. City inside Portland, population 752
4. Shipbuilding city destroyed by flood
6. Annexed by Portland in 1915
8. Former town for meatpacking company
9. Amusement park site from 1907 to 1929
10. "The truth will set you free"
11. Home of the world's second-oldest chiropractic doctoral program
12. Over eighty miles of recreational trails
13. Former home of 24 Hour Church of Elvis
14. Formerly Holyrood
17. Where Katherine Dunn got the idea for *Geek Love*
19. Home to Broadway Books
20. Part of Boring Lava Field
24. From warehouses to art galleries
25. Twelve-mile stretch, a former Superfund site
29. Home to Lone Fir Cemetery
30. Location of first Stumptown Roasters
31. A creek runs through it
32. Designed as a promenade and a firebreak
34. Neon martini glass
35. Home of Big Pink
38. Submarine docked here
39. Former rail corridor
41. Former site of a highway
44. Home of the Timbers and the Thorns
46. Not the one where Hendrix and Joplin played
47. Home of the Pickles

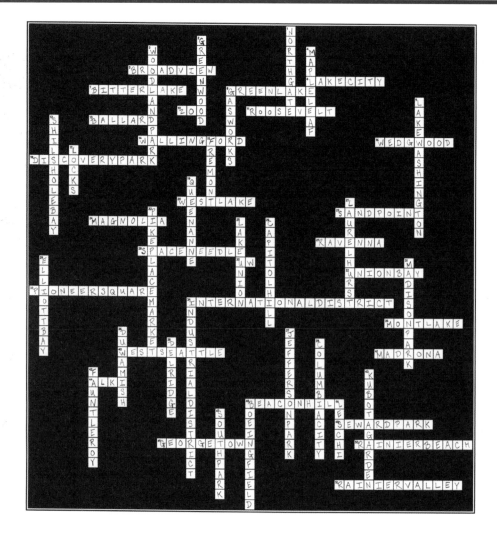

ACROSS

5 Where early farmers floated their wares to Seattle
6 Incorporated as the fourth-largest community in Washington in 1949
7 Home to Playland amusement park from 1930 to 1961
8 Where the Seafair Milk Carton Derby is held
10 Second to Bronx Zoo for number of exhibit awards won
11 Named for Teddy, not Franklin
13 Buy lutefisk here
14 Location of first Dick's Drive-In
16 Area often misspelled
18 Largest city park
20 By area, the city's smallest neighborhood
23 Location of Sound Garden
24 Connected by three bridges to the city
27 Named after Italian city
28 Built for the 1962 World's Fair
30 Campus used for the 1909 Alaska-Yukon-Pacific Exposition
32 Location of Seattle's largest garbage dump for forty years
33 Birthplace of modern Seattle
34 Chinatown, Japantown, Little Saigon
35 Hosts opening day of boating season
40 Wes C. Addle (Eddie Vedder) lives here
41 Named after common tree in the area
44 Original white settlement in Seattle
45 Named after Beacon Hill in Boston, Massachusetts
48 Named for US Secretary of State William Seward, of Seward's Folly
49 Hat 'n' Boots Park
50 This high school regularly produces NBA players
51 The most diverse neighborhood in the country in 2010

DOWN

1 First mall in the United States, kind of
2 Once mostly marshland
3 Warren G. Harding spoke here six days before dying in San Francisco
4 National Neighborhood of the Year in 1986
8 Contains structures from a former coal gasification plant
9 Level dropped nine feet thanks to the Montlake Cut
12 Duwamish for threading a needle
15 Lenin statue moved here from Slovakia
17 Carries more boat traffic than any other lock in the United States
19 Counterbalance was used for electric trolleys climbing this hill
21 Onetime home of Bill Gates, Gary Larson, Duff McKagan
22 Place with flying fish
25 Tom Hanks was sleepless there
26 Site of Coryell Court Apartments in *Singles*
29 Lent its name to a noted bookstore
31 First baseball in Seattle
36 Indigenous people of Seattle
37 Home course of Fred Couples
38 Has an operative steel plant
39 Hitt Fireworks Company
42 Neighborhood, cove, creek, park, ferry
34 Sears to Starbucks
43 Started in 1927 by Japanese emigrant Fujitaro Kubota
45 City's primary airport from 1928 until 1944
46 Named for chief of the Nisqually nation
47 Riverfront neighborhood not named after irreverent cartoon series

WORD SEARCH

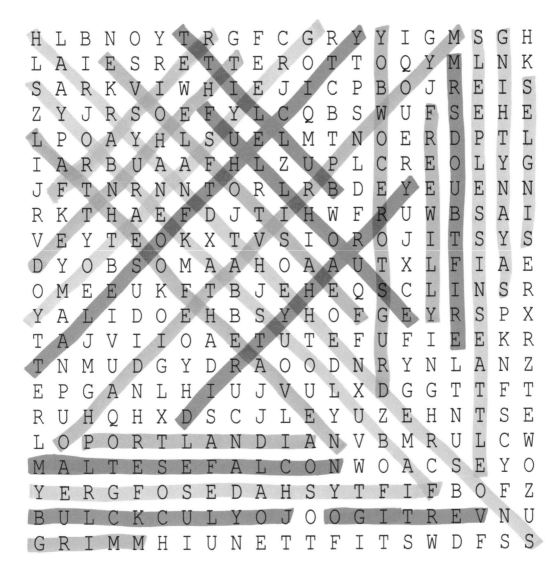

```
H L B N O Y T R G F C G R Y Y I G M S G H
L A I E S R E T T E R O T T O Q Y M L N K
S A R K V I W H I E J I C P B O J R E I S
Z Y J R S O E F Y L C Q B S W U F S E H E
L P O A Y H L S U E L M T N O E R D P T L
I A R B U A A F H L Z U P L C R E O L Y G
J F T N R N N T O R L R B D E Y E U E N N
R K T H A E F D J T I H W F R U W B S A I
V E Y T E O K X T V S I O R O J I T S Y S
D Y O B S O M A A H O A A U T X L F I A E
O M E E U K F T B J E H E Q S C L I N S R
Y A L I D O E H B S Y H O F G E Y R S P X
T A J V I I O A E T U T E F U F I E E K R
T N M U D G Y D R A O O D N R Y N L A N Z
E P G A N L H I U J V U L X D G G T T F T
R U H Q H X D S C J L E Y U Z E H N T S E
L O P O R T L A N D I A N V B M R U L C W
M A L T E S E F A L C O N W O A C S E Y O
Y E R G F O S E D A H S Y T F I F B O F Z
B U L C K C U L Y O J O O G I T R E V N U
G R I M M H I U N E T T F I T S W D F S S
```

SAN FRANCISCO

VERTIGO
MALTESE FALCON
BULLITT
MRS. DOUBTFIRE
DIRTY HARRY
JOY LUCK CLUB
FULL HOUSE
TALES OF THE CITY

PORTLAND

MY OWN PRIVATE IDAHO
FEAST OF LOVE
LATHE OF HEAVEN
FREE WILLY
THE HUNTED
DRUGSTORE COWBOY
GRIMM
PORTLANDIA

SEATTLE

SLEEPLESS IN SEATTLE
FIFTY SHADES OF GREY
SINGLES
SAY ANYTHING
FABULOUS BAKER BOYS
HARRY AND THE HENDERSONS
GREY'S ANATOMY
FRASIER

AUTHORS' NOTE

This atlas is a collaboration that includes the cartographic, textual, graphic, and photographic contributions of more than forty people with the input and involvement of dozens more. Hunter Shobe, David Banis, and Zuriel van Belle collectively worked on the conceptualization and design of every page. David and Hunter managed the production of the content and shepherded the maps, text, graphics, and photographs to completion. The often collaborative design of the maps belongs to the cartographers, the design of the essays belongs to the writers, and the design of the graphics belongs to the illustrators.

ACKNOWLEDGMENTS

Hunter, David, and Zuriel thank the following people:

All of our contributors for their great ideas, hard work, dedication, and camaraderie. Jonathan van Belle and Kelly Neeley-Brown for their insightful comments and suggestions on drafts of the book. Devin Busby and Mary Hansen at the Portland Archives, Christina Moretta at the San Francisco History Center, and Julie Irick at the Seattle Municipal Archives for their generous help with historical photographs. Alan Canterbury, Michael Henniger, and Donald Nelson for kindly allowing us to use their historical photographs. Tim Berkley, Michael Endicott, Rachel Escoto, and C. Bruce Forster for allowing us to use their photographs. Abbey Gaterud, Carl Abbott, and Chet Orloff for their ongoing advice and support. Tomoya Hanibuchi for the continuing exchange of ideas about cultural atlases and cultural cartography. The PSU Geography Department classes Urban Landscapes Fall 2017, Maps and Society Winter 2018, and Research Methods in Human Geography Spring 2018, for their brainstorming, research, and energy. The Portland Community College Geography classes Local Landscapes Fall 2018, Applications in GIS in Spring 2019, and their instructor Christina Friedle for their ideas and fresh perspectives. Portland State University for the Faculty Development Grant that helped support our work during 2017–2018. Liza Brice-Dahmen, Anna Goldstein, Alison Keefe, Gary Luke, Jill Saginario, and Jennifer Worick, and the whole team at Sasquatch for believing in our project.

CONTRIBUTIONS AND KEY DATA SOURCES

Throughout this book, we use city and regional map data obtained from the following sources:
- San Francisco Open Data: DataSF.org
- Seattle Open Data: data.seattle.gov
- King County Open Data: data.kingcounty.gov
- PortlandMaps—Open Data: GIS-PDX.OpenData.ArcGIS.com
- Metro Regional Land Information System Discovery: rlisdiscovery.oregonmetro.gov

INTRODUCTION: UPPER LEFT
CARTOGRAPHY: Zuriel van Belle
TEXT: Hunter Shobe, Zuriel van Belle, and Jonathan van Belle
RESEARCH: Hunter Shobe and Zuriel van Belle
SPECIAL THANKS TO: Nick Martinelli
KEY DATA SOURCES:
DeLeon, Richard Edward. *Left Coast City: Progressive Politics in San Francisco, 1975–1991.* Lawrence: University Press of Kansas, 1992.
Gregory, James N. "Seattle's Left Coast Formula." *Dissent,* winter 2015.
Safire, William. "On Language." *New York Times Magazine,* October 1, 2000.

BACK IN THE DAY
CARTOGRAPHY: David Banis, Jon Franczyk (Seattle), Gabriel Rousseau (San Francisco), and Joseph Bard (Portland)
TEXT: Hunter Shobe
RESEARCH: David Banis and Hunter Shobe
KEY DATA SOURCES:
Abbott, Carl. *Greater Portland: Urban Life and Landscape in the Pacific Northwest.* Philadelphia: University of Pennsylvania Press, 2001.
———. *Portland in Three Centuries: The Place and the People.* Corvallis: Oregon State University Press, 2011.
Crowley, Walt. "Seattle—A Brief History of Its Founding." HistoryLink.org, August 31, 1998.
———. "Seattle—Thumbnail History." HistoryLink.org, September 26, 2006.
DuwamishTribe.org. "Chief Si'ahl." Accessed August 24, 2019.
Friends of Mt. Tabor Park. "Mt. Tabor's Volcanic History." Accessed August 24, 2019.
Hooper, Bernadette. *San Francisco's Mission District.* Charleston: Arcadia Publishing, 2006.
Issel, William, and Robert W. Cherney. *San Francisco, 1865–1932: Politics, Power, and Urban Development.* Berkeley: University of California Press, 1986.
Lange, Greg. "Seattle's Denny Regrade Is Completed after 32 Years on December 10, 1930." HistoryLink.org, January 16, 1999. Accessed August 31, 2019.

Richards, Rand. *Historic San Francisco: A Concise History and Guide.* San Francisco: Heritage House Publishers, 1997.
Sale, Roger. *Seattle: Past to Present.* Seattle: University of Washington Press, 1976.
Seattle Municipal Archives. "Brief History of Seattle." Accessed August 24, 2019.

WHAT IS A METRO AREA?
CARTOGRAPHY: Zuriel van Belle and Alicia Milligan
TEXT: Hunter Shobe and David Banis
RESEARCH: Zuriel van Belle and David Banis

SOMEWHERE IN THE NEIGHBORHOOD OF
CARTOGRAPHY: Randy Morris and David Banis
TEXT: Hunter Shobe and David Banis
RESEARCH: David Banis, Randy Morris, and Hunter Shobe
KEY DATA SOURCES:
Crowley, Walt. "Seattle Neighborhoods—Past, Present, Future." HistoryLink.org, January 4, 2005.

PROFILES IN DEMOGRAPHY
CARTOGRAPHY: David Banis and Randy Morris
TEXT: Hunter Shobe and David Banis
RESEARCH: David Banis and Matthew Gregg
KEY DATA SOURCES:
US Census Bureau. "2017 American Community Survey: Five-Year Estimates." Factfinder.census.gov.
Zillow. Data: Home Values. Accessed August 24, 2019. https://www.zillow.com/research/data/.

GLOBAL POSITIONING
CARTOGRAPHY: David Banis
TEXT: Hunter Shobe
RESEARCH: Hunter Shobe, David Banis, and Matthew Gregg
PHOTOGRAPHY: Patoir, Armand. *Je veux rentrer . . . à PAF* (Port-aux-Français, Kerguelen). August 26, 2018. Accessed August 29, 2019. Retrieved from Wikimedia Commons. Commons.Wikimedia.org.
KEY DATA SOURCES:
Buhariwalla, Colin. "Episode 2—Greetings from Port-aux-Français." OceanTrackingNetwork.org, February 17, 2016.
Wikitravel.org. "Kerguelen" entry. Accessed August 24, 2018.

I: URBAN LANDSCAPES

BREAKING THE GRID
CARTOGRAPHY: Alicia Milligan, Lauren McKinney-Wise, and David Banis
TEXT: Hunter Shobe

RESEARCH: David Banis and Alicia Milligan

INSPIRATION: City block maps inspired by the work of artist Armelle Caron.

KEY DATA SOURCES:

Abbott, Carl. "Ladd's Addition." The Oregon Encyclopedia. Accessed September 1, 2019.

Argay Terrace Neighborhood Association. "Argay Terrace History." ArgayTerrace.org, no date. Accessed September 1, 2019.

Asimov, Nanette. "Dorothy Adams Dies: Broke Race Restriction on Homeowners in SF." SFGate.com, January 13, 2015. Accessed September 1, 2019.

Fiset, Louis. "Seattle Neighborhoods: Haller Lake—Thumbnail History." HistoryLink.org, July 22, 2001.

OutsideLands.org. "Ingleside Racetrack." Accessed September 1, 2019.

———. "Ingleside Terraces." Accessed September 1, 2019.

———. "Westwood Park." Accessed September 1, 2019.

Parkmerced Vision. "A Place Like No Other." Accessed September 1, 2019.

Rochester, Junius. "Seattle Neighborhoods: Laurelhurst—Thumbnail History." HistoryLink.org, June 9, 2001.

Westwood Park Association. "Westwood Park." Accessed September 1, 2019.

Wilma, David. "Seattle Neighborhoods: View Ridge—Thumbnail History." HistoryLink.org, July 24, 2001. Accessed September 1, 2019.

Works, Martha. "Laurelhurst." The Oregon Encyclopedia. Accessed September 1, 2019.

WATER UNDER THE BRIDGE

CARTOGRAPHY: Lauren McKinney-Wise

GRAPHICS: Zuriel van Belle, Lauren McKinney-Wise, and David Banis

TEXT: Hunter Shobe

RESEARCH: Zuriel van Belle and Hunter Shobe

KEY DATA SOURCES:

Hallman Jr., Tom. "Steel Bridge: An Engineering Marvel That Keeps Up with the Times: 'Spanning Oregon.'" *Oregonian*, November 28, 2015. Updated January 9, 2019.

Multnomah County. "Willamette River Bridges." Accessed September 24, 2019.

National Oceanic and Atmospheric Administration. "Office of Coast Survey: Electronic Charts." Accessed September 24, 2019.

Portland Yacht Club. "Portland Bridges." Accessed September 24, 2019.

Seattle Department of Transportation. "Bridges." Accessed September 24, 2019.

Sigmund, Pete. "The Golden Gate: 'The Bridge That Couldn't Be Built.'" ConstructionEquipmentGuide.com, June 2, 2006.

Washington State Department of Transportation. "SR 520 Bridge Replacement and HOV Program—Floating Bridge Facts." Accessed August 24, 2019.

West Seattle Connection. *World's Only Hydraulically Operated Double-Leaf Concrete Swing Bridge*. City of Seattle Engineering Department. Accessed August 7, 2019.

TRAFFIC LIGHT TAPESTRY

CARTOGRAPHY: Sachi Arakawa and Alicia Milligan

TEXT: Sachi Arakawa

RESEARCH: Sachi Arakawa

NO WASTE OF SPACE: ALLEYS

CARTOGRAPHY: David Banis and Alicia Milligan

TEXT: Sachi Arakawa and Hunter Shobe

RESEARCH: Sachi Arakawa, David Banis, and Hunter Shobe

PHOTOGRAPHY:

Banis, David. *Alley in Concordia*. Portland, 2019.

———. *Alley in Ladd's Addition*, Portland, 2019.

Shobe, Hunter. *Alley in the Sunset District*. San Francisco, 2017.

———. *Upper Post Alley (Behind Stewart House Looking North)*. Seattle, 2019.

Upper Post Alley (Behind Stewart House Looking North). Color photograph, ca. 1978. Seattle Municipal Archives Photographs Collection, Pike Place Market Visual Images and Audiotapes. Record series 1628-02, item number 204.6C.

KEY DATA SOURCES:

Abbott, Carl. "Ladd's Addition." The Oregon Encyclopedia. Accessed August 7, 2019.

Alleys and stairs data derived from OpenStreetMap data downloaded from Download.Geofabrik.de.

Canin Associates. "Alleys in Urban Design: History and Application." Accessed August 7, 2017.

Drescher, Timothy. "Street Subversion: The Political Geography of Murals and Graffiti." In *Reclaiming San Francisco: History, Politics, Culture*, 231–245. San Francisco: City Lights, 1998.

Grant, Benjamin. "Why We Love Alleys." SPUR.org. Accessed August 7, 2019.

Hale, Jamie. "The Alley Sweeper, Portland's Back Alley Motorcycle Ride, Is Swept Underground." *Oregonian*, March 12, 2015. Accessed August 7, 2019.

Keeling, Brock. "Chinatown Alleyways: A Tour of Our Favorites." Curbed San Francisco, February 1, 2019. Accessed August 7, 2019.

Otárola, Miguel. "Seattle's Alley Get a Face-Lift." *Seattle Times*, July 9, 2015. Updated July 10, 2015. Accessed August 7, 2019.

Portland Alley Project. Accessed August 7, 2019. DSDauphin.WordPress.com.

Seattle Department of Transportation. "Alleys." Accessed August 7, 2017.

———. "Canton, Nord, and Pioneer Passage Alley Improvement Project." Accessed August 7, 2017.

Veterans Alley. Accessed August 7, 2019. VetsAlley.org.

PUBLIC SPACES: THE COMMONS, THE WATERFRONT, AND THE EMBARCADERO

CARTOGRAPHY: Gabriel Rousseau

TEXT: Sarah Mercurio

RESEARCH: David Banis, Gabriel Rousseau, and Sarah Mercurio

PHOTOGRAPHY:

Shobe, Hunter. *South Lake Union Photos*. Seattle, 2019.

Steel Bridge Approaches from Front Ave. and Harbor Dr. Black and white photograph, ca. 1949. City of Portland (OR) Archives, Public Works Administration (Archival), record number AP/5440, PARC accession A2005-001.634.

KEY DATA SOURCES:

American Planning Association. "Governor Tom McCall Waterfront Park: Portland, Oregon." Accessed August 7, 2019. Planning.org.

Becker, Paula. "Seattle Voters Reject the Seattle Commons Levy on September 19, 1995." HistoryLink.org, August 8, 2007. Accessed August 7, 2019.

Berger, Knute. "Visions of the Seattle That Could Have Been." CityLab, February 6, 2013. Accessed August 7, 2019.

Carlsson, Chris. "The Freeway Revolt." FoundSF.org, no date. Accessed August 7, 2019.

City of Portland. "Waterfront Park Master Plan." Portland Parks and Recreation. Accessed August 7, 2019. Portlandoregon.gov.

Commons map re-created from *Seattle Commons Draft Plan*, 1995, Seattle Municipal Archives Digital Collection, Record series 9910-03. Identifier 339. Created 1995.

Historic highways derived from 1:24000-scale USGS topographic maps (Portland, 1954; San Francisco, 1969).

Jenner, Michael Anthony. "The Origin of Portland, Oregon's Waterfront Park: A Paradigm Shift in City Planning (1967–1978)." Thesis, March 4, 2004. Portland State University. Accessed August 7, 2019.

Keenan, Edward. "San Francisco's Waterfront Freeway Was Removed 25 Years Ago: No One Misses It." *The Star*, June 5, 2015. Accessed August 7, 2019.

Lloyd, M. "Portland's Harbor Drive Was an Urban Development Landmark, Before Going Away." *Oregonian*, May 14, 2012. Accessed August 7, 2019.

No More Freeway Expansions. "Our Advocacy." Accessed August 7, 2019.

Scigliano, Eric. "The Seattles That Might Have Been." *Seattle Times' Pacific NW Magazine*, June 7, 2018. Accessed August 7, 2019.

Stein, Mark A., and Kaufman, Norma. "Future of Embarcadero Freeway Divides San Francisco." *Los Angeles Times*, April 13, 1990. Accessed August 7, 2019.

SKYLINES

GRAPHICS: Maggie Burant

TEXT: Hunter Shobe

RESEARCH: Maggie Burant and Hunter Shobe

GRAVEYARD SHIFT

CARTOGRAPHY: Alicia Milligan and David Banis

TEXT: Hunter Shobe

RESEARCH: Hunter Shobe and David Banis

PHOTOGRAPHY:

Shobe, Hunter. *San Francisco National Cemetery*. San Francisco, 2019.

KEY DATA SOURCES:

Anderson, Rick. "No Stone Unturned: One Man's Lonely Battle to Save the Graveyard City Hall Would Rather Forget." *Seattle Weekly*, October 9, 2006. Accessed August 7, 2019.

Angotti, Laura. "Seattle's Denny Hotel Cemetery." HistoryLink.org, November 2, 1998. Accessed August 7, 2019.

———. "Seattle Cemetery." HistoryLink.org, March 16, 1999. Accessed August 7, 2019.

Branch, John. "The Town of Colma, Where San Francisco's Dead Live." *New York Times*, February 5, 2016. Accessed August 7, 2019.

Brooks, Jon. "Why Are There So Many Dead People in Colma? And So Few in San Francisco?" KQED, October 26, 2017. Accessed August 7, 2019.

FoundSF.org. "Old Cemeteries in the City." Accessed August 24, 2019.

Metro. "Columbia Pioneer Cemetery." Accessed August 7, 2109. Oregonmetro.gov.

———. "Lone Fir Cemetery." Accessed August 7, 2019. Oregonmetro.gov.

Park, Chris C. *Sacred Worlds: An Introduction to Geography and Religion*. New York: Routledge, 1994.

Plan for Denny Park, 1884, Part of Seattle Cemetery Removal and Reburial Register, Seattle Municipal Archives, 5801_01_001_001_001.

Proctor, William A. *City Planner Report: Location, Regulation, and Removal of Cemeteries in the City and County of San Francisco*. City and County of San Francisco. SF Genealogy.com, 1950. Accessed August 7, 2019.

ON THE WATERFRONT

CARTOGRAPHY: Kristin Sellers

TEXT: Hunter Shobe

RESEARCH: Kristin Sellers and David Banis

URBAN MOSAICS

CARTOGRAPHY: Corinna Kimball-Brown

PHOTOGRAPHY:

Banis, David. *Green Photos of Portland*. Portland, 2014–2015.

Kimball-Brown, Corinna. *Red Photos of San Francisco*. San Francisco, 2019.

Shobe, Cielo. *Red Photos of San Francisco*. San Francisco, 2019.

Shobe, Hunter. *Blue Photos of Seattle*. Seattle, 2019.

———. *Green Photos of Portland*. Portland, 2019.

———. *Red Photos of San Francisco*. San Francisco, 2018–2019.

Shobe, Soraya. *Red Photos of San Francisco*. San Francisco, 2019.

TEXT: Hunter Shobe

II: NATURE AND THE CITY

WHAT ARE YOU ON? LAND USE IN THE METRO AREA
CARTOGRAPHY: Zuriel van Belle
GRAPHICS: Zuriel van Belle
TEXT: Zuriel van Belle
RESEARCH: Zuriel van Belle
KEY DATA SOURCES:
Multi-Resolution Land Characteristics Consortium. "National Land Cover Database: 2011." Mrlc.gov.

CREATURES OF HABITAT
CARTOGRAPHY: Zuriel van Belle
GRAPHICS: Kezia Rasmussen
TEXT: Zuriel van Belle
RESEARCH: Zuriel van Belle and Kezia Rasmussen
INSPIRATION: The wildlife map was inspired by a 1955 Russian classroom roll-down map of uncertain origin.

A CAPTIVE AUDIENCE: ZOOS
CARTOGRAPHY: Zuriel van Belle
GRAPHICS: Zuriel van Belle
TEXT: Zuriel van Belle
RESEARCH: Zuriel van Belle
PHOTOGRAPHY:
City (Washington) Park Bear Pit. Black and white photograph, ca. 1935. City of Portland (OR) Archives, City Auditor—Archives and Records Management, record number AP/4896, PARC accession A2004-002.857.
Woodland Park Zoo Animals (Deer). Silver gel print, ca. 1914. Seattle Municipal Archives Photograph Collection, Don Sherwood Parks History Collection, record series 5801-01, item number 30826.
SPECIAL THANKS TO: the Oregon Zoo, the San Francisco Zoo, and the Woodland Park Zoo for generously providing data.
KEY DATA SOURCES:
Elefterakis, Anne. Education Department, San Francisco Zoological Society. Personal communication, 2018.
Hanson, Elizabeth. *Animal Attractions*. Princeton, New Jersey: Princeton University Press, 2002.

TRAILS AS TRANSIT MAPS
CARTOGRAPHY: Geoff Gibson
TEXT: Geoff Gibson
RESEARCH: Geoff Gibson
KEY DATA SOURCES:
BayTrail. "The San Francisco Bay Trail." Accessed August 9, 2019. BayTrail.org.
King County Parks. "Regional Trails in King County." Accessed August 9, 2019. KingCounty.Maps.ArcGIS.com.
———. "Trails." Accessed August 9, 2019. Kingcounty.gov.
Metro. "Regional Trails System: 2018." Accessed August 9, 2019. Oregonmetro.gov.
———. "Regional Trails System Plan." Accessed August 9, 2019. Oregonmetro.gov.

Metropolitan Transportation Commission. "Plans + Projects: San Francisco Bay Trail." Accessed August 9, 2019. Mtc.ca.gov.
TrailLink. "San Francisco, CA Trails and Maps." Accessed August 9, 2019.

DRINKING WATER
CARTOGRAPHY: David Banis and Sachi Arakawa
GRAPHICS: Jessica Sullivan
TEXT: Sachi Arakawa and Hunter Shobe
RESEARCH: Sachi Arakawa and David Banis
KEY DATA SOURCES:
HetchHetchy.org. "Restore Hetch Hetchy." Accessed September 1, 2019.
Ingalls, Libby. "Spring Valley Water Company." FoundSF.org, no date. Accessed September 1, 2019.
Kelleher, Susan. "Major Do-over for Two Seattle Reservoirs." *Seattle Times*, July 17, 2009. Accessed September 1, 2019.
Kost, Ryan. "Portland Fluoride: For the Fourth Time Since 1956, Portland Voters Reject Fluoridation." *Oregonian*, January 10, 2019. Accessed September 1, 2019.
National Park Service. "Remember Hetch Hetchy: The Raker Act and the Evolution of the National Park Idea." Posted December 20, 2013. Accessed September 1, 2019. Nps.gov.
Oldham, Kit. "Seattle Residents Receive Cedar River Water for the First Time on January 10, 1901." HistoryLink.org, October 17, 2014. Accessed September 1, 2019.
Portland Water Bureau. "History." Accessed September 1, 2019. Portlandoregon.gov.
———. "Uncovered Reservoirs." Accessed September 1, 2019. Portlandoregon.gov.
Rayman, Noah. "Portland Dumps 38 Million Gallons of Water After Man Pees in Reservoir." *Time*, April 17, 2014. Accessed September 1, 2019.
San Francisco Water Power Sewer, San Francisco Public Utilities Commission. "Overview: Serving 2.7 Million Residential, Commercial and Industrial Customers." Accessed September 1, 2019. SFWater.org.
Seattle Municipal Archives. "Water System." Accessed September 1, 2019. Seattle.gov.
Seattle Public Utilities. "Protected Watersheds." Accessed September 1, 2019. Seattle.gov.
———. "Reservoir Covering." Accessed September 1, 2019. Seattle.gov.
US Environmental Protection Agency. "Drinking Water Requirements for States and Public Water Systems." Accessed September 1, 2019. Epa.gov.

CONCRETE JUNGLE
CARTOGRAPHY: Justin Sherrill and Kristin Sellers
GRAPHICS: Stephanie Sun
TEXT: Hunter Shobe and David Banis
RESEARCH: Justin Sherrill, Matt Downs, and David Banis
KEY DATA SOURCES:
PortlandMaps—Open Data. GIS-PDX.OpenData.ArcGIS.com.
San Francisco Open Data. DataSF.org.
Seattle Open Data. Data.seattle.gov.

NATURAL DISASTERS

CARTOGRAPHY: Joseph Bard

TEXT: Sarah Mercurio, Joseph Bard, and Hunter Shobe

RESEARCH: Sarah Mercurio, Joseph Bard, and David Banis

PHOTOGRAPHY:

Flooded intersection of SW Third Ave. and SW Washington St. Black and white photograph, June 30, 1894. City of Portland (OR) Archives, City Auditor—Archives and Records Management, record number AP/4919, PARC accession A2004-002.699.

KEY DATA SOURCES:

City of Portland Archives. *Lewis and Dryden's High Water Map of Portland: 1894 Flood*. Map. Record A2004-002. Portlandoregon.gov.

City of Seattle Open Data Portal. "Unreinforced Masonry Buildings." September 11, 2019. Data.seattle.gov.

DNR GIS Data. Shaking Amplification Polygons from Washington State Department of Natural Resources. Accessed September 25, 2019. Data-WADNR.OpenData. ArcGIS.com.

Oregon Military Department. "Emergency Management Resources: Cascadia Rising 2016." Oregon Office of Emergency Management. Accessed August 9, 2019. Oregon.gov.

Ott, Jennifer. "Vanport Flood Begins on Columbia River on May 30, 1948." HistoryLink.org, August 30, 2013. Accessed August 9, 2019.

San Francisco Open Data. "Tall Building Inventory." October 2, 2018. Seismic Hazard Zones/Liquefaction zones. Accessed September 25, 2019. DataSF.org.

——. "San Francisco Seismic Hazard Zones." July 12, 2016. Accessed September 25, 2019. DataSF.org.

San Francisco Planning. "Sea Level Rise Vulnerability and Consequences Assessment." Accessed August 9, 2019. SFPlanning.org.

Sulek, Julia Prodis, Lisa M. Krieger, and Mark Gomez. "Russian River Flooding Swamps Two Dozen Towns." *Mercury News*, February 27, 2019. Accessed August 9, 2019.

US Army, Corps of Engineers, Pacific Ocean Division. *Map of San Francisco, California*. Map. Record B22254311. UC Berkeley Earth Sciences and Map Library, Digitized. Accessed September 25, 2019.

US Environmental Protection Agency. "Air Data: Air Quality Data Collected at Outdoor Monitors Across the US." Epa.gov.

US Geological Survey. "Faults: Quaternary Fault and Fold Database of the United States." Earthquake.usgs.gov /hazards/qfaults.

——. "Earthquakes." Usgs.gov.

Washington Military Department. "Looking at Successes of Cascadia Rising and Preparing for Our Next Big Exercise." June 7, 2018. Accessed August 9, 2019. Mil.wa.gov.

Washington State Department of Natural Resources. "Earthquakes and Faults." https://geologyportal.dnr.wa.gov/.

SASQUATCH MEANS BUSINESS

CARTOGRAPHY: Zuriel van Belle

GRAPHICS: Maggie Burant

TEXT: Hunter Shobe and Zuriel van Belle

RESEARCH: Zuriel van Belle

KEY DATA SOURCES:

IMDb.com. "Finding Bigfoot." Accessed August 9, 2019.

——. "Harry and the Hendersons." Accessed August 9, 2019.

——. "Patterson-Gimlin Film." Accessed August 9, 2019.

Shah, Haleema. "The Scientist Grover Krantz Risked It All . . . Chasing Bigfoot." Smithsonian.com, October 31, 2018. Accessed August 9, 2019.

Thomas, Nicki. "Sasquatch." *The Canadian Encyclopedia*, January 21, 2007. Accessed August 9, 2019.

III: SOCIAL RELATIONS

WE ARE FAMILY: SISTER CITIES

CARTOGRAPHY: David Banis

TEXT: Hunter Shobe

RESEARCH: Hunter Shobe and David Banis

PHOTOGRAPHY:

Banis, David. *Terry Schrunk Plaza*. Portland, 2014.

Kimball-Brown, Corinna. *Hallidie Plaza*. San Francisco, 2019.

Shobe, Hunter. *Daejeon Park*. Seattle, 2019.

KEY DATA SOURCES:

City of Portland. "Sister Cities: About." Accessed August 9, 2019. Portlandoregon.gov.

Office of Intergovernmental Associations. "Sister Cities: About Seattle Sister Cities." Accessed August 31, 2018 and August 9, 2019. Seattle.gov.

Portland-Bologna Sister City Association. "Home." Accessed August 9, 2019. Portland-Bologna.org.

San Francisco Office of Economic and Workforce Development. "San Francisco Sister Cities." Accessed August 9, 2019. OEWD.org.

Seattle Parks and Recreation. "Kobe Terrace." Accessed August 29, 2019. Seattle.gov.

DEEP PURPLE: VOTING IN THE UPPER LEFT

CARTOGRAPHY: David Banis and Matthew Gregg

GRAPHICS: Matthew Gregg and David Banis

TEXT: Hunter Shobe, David Banis, and Matthew Gregg

RESEARCH: David Banis, Matthew Gregg, and Hunter Shobe

KEY DATA SOURCES:

King County Elections Department. "King County Elections." Kingcounty.gov.

——. "Past Elections." Kingcounty.gov.

Oregon Secretary of State Elections Division. "Oregon Election Historical Results and Data." Sos.oregon.gov.

Portland State University. "Who Votes for Mayor?" Accessed August 19, 2019. WhoVotesforMayor.org.

Washington Secretary of State Elections Division. "Data and Research." Sos.wa.gov.

BLUE NOTES: LOST JAZZ CLUBS

CARTOGRAPHY: Zuriel van Belle
TEXT: Hunter Shobe
RESEARCH: Hunter Shobe
PHOTOGRAPHY:
Hanibuchi, Tomoya. *Ernestine Anderson Sign on Jackson Street.* Seattle, 2019.
KEY DATA SOURCES:
Armbruster, Kurt E. *Before Seattle Rocked: A City and Its Music.* Seattle: University of Washington Press, 2011.
Bern, Medea Isphording. *San Francisco Jazz.* Charleston: Arcadia Publishing, 2014.
Blecha, Peter. "Cecil Young's Bebop Jazz: Seattle 1951." *Northwest Music Archives,* 2015. Accessed August 9, 2019. NW-Music-Archives.Blogspot.com.
Darroch, Lynn. *Rhythm in the Rain: Jazz in the Pacific Northwest.* Portland: Ooligan Press, 2016.
De Barros, Paul. *Jackson Street after Hours: The Roots of Jazz in Seattle.* Seattle: Sasquatch Books, 1993.
Dietsche, Robert. *Jumptown: The Golden Years of Portland Jazz, 1942–1957.* Corvallis: Oregon State University Press, 2005.
Faltys-Burr, Kaegan. "Jazz on Jackson Street: The Birth of a Multiracial Music Community in Seattle." Civil Rights and Labor History Consortium, University of Washington, 2010. Accessed August 9, 2019.
Metcalfe, Rochelle. "From Jimbo's to Yoshi's." *New Fillmore,* 2011.
Pepin, Elizabeth, and Lewis Watts. *Harlem of the West: The San Francisco Fillmore Jazz Era.* San Francisco: Chronicle Books, 2006.

ENCAMPMENTS

CARTOGRAPHY: David Banis and Alicia Milligan
TEXT: Hunter Shobe and David Banis
RESEARCH: Krystle Harrell, David Banis, and Hunter Shobe
KEY DATA SOURCES:
Archibald, Ashley. "KIRO Pulls Map of Homeless Encampments." *Real Change News,* May 15, 2017. Accessed September 2, 2019.
City of Seattle. "Homeless Response—Unauthorized Encampments Emphasis Map." Accessed September 9, 2019. Seattle.gov.
Department of Homelessness and Supportive Housing. "San Francisco Homeless Point in Time Count Reports: 2019 Executive Summary." Accessed September 9, 2019. Hsh.sfgov.org.
Harrell, Krystle. "Homelessness in Portland, Oregon: An Analysis of Homeless Campsite Spatial Patterns and Spatial Relationships." PDXScholar, May 2019. Accessed September 2, 2019. PDXScholar.Library.PDX.edu.
Portland's One Point of Contact Campsite Reports data from 2018 courtesy of Nikki Harrell.

THERE GOES THE GAYBORHOOD

CARTOGRAPHY: Jonathan Leeper and Zuriel van Belle
TEXT: Jonathan Leeper
RESEARCH: Jonathan Leeper
INSPIRATION: Our title for these pages was inspired by Amin Ghaziani's book *There Goes the Gayborhood.* New Jersey: Princeton University Press, 2015.
PHOTOGRAPHY:
Shobe, Hunter. *Rainbow Crosswalk, The Castro.* San Francisco, 2017.
KEY DATA SOURCES:
Bowles, Nellie, and Sam Levin. "San Francisco's Tech Bros Told: Quit Changing the Gayborhood." *Guardian,* February 2, 2016. Accessed August 10, 2019.
Brown, Michael P. *Claiming Space: Seattle's Lesbian and Gay Historical Geography* (maps). Seattle: Northwest Lesbian and Gay History Museum Project, 2004. Accessed August 10, 2019.
Carder, Justin. "Finding 'Seattle's Shifting Queer Geographies' in the More Recent History of Capitol Hill." *Capitol Hill Seattle Blog,* February 28, 2018. Accessed August 20, 2019. CapitolHillSeattle.com.
Gay and Lesbian Archives of the Pacific Northwest. "1999 Portland Gay History Walking Tour." Accessed August 10, 2019. GLAPN.org.
Kohl, David Grant. *A Curious and Peculiar People.* Portland: Spirit Press, 2006.
Mandel, Eric. "Pike/Pine's Booming Bar Scene Comes at a Price for Gay Nightlife." *Capitol Hill Seattle Blog,* April 4, 2014. Accessed August 20, 2019. CapitolHillSeattle.com.
McKenna, Kevin, and Michael Aguirre. "A Brief History of LGBTQ Activism in Seattle." Seattle Civil Rights and Labor History Project. University of Washington. Accessed August 31, 2019. Depts.Washington.edu.
SF Gay History. Accessed August 10, 2019. SFGayHistory.com.
Stafford, Zach. "Violence in Capitol Hill: Is This the End of the Line for Seattle's Gay Neighborhood?" *Guardian,* February 23, 2016. Accessed August 10, 2019.

UPRISING

CARTOGRAPHY: Randy Morris and David Banis
GRAPHICS: Geoff Gibson and Jessica Sullivan
TEXT: Hunter Shobe, David Banis, and Geoff Gibson
RESEARCH: David Banis, Geoff Gibson, Julie Wilcox, Hunter Shobe, and Randy Morris

UNKNOWN MEASURES

CARTOGRAPHY: Justin Sherrill
TEXT: Hunter Shobe
RESEARCH: Justin Sherrill and Hunter Shobe
KEY DATA SOURCES:
Christiansen, Jen. "Pop Culture Pulsar: Origin Story of Joy Division's Unknown Pleasures Album Cover." *Scientific American,* February 18, 2015. Accessed August 10, 2019. Blogs.ScientificAmerican.com.
George, Cassidy. "Peter Saville's Joy Division Artwork Is Still Influencing Fashion 40 Years On." Highsnobiety.com, June 18, 2019. Accessed August 10, 2019.

MORE THAN WORDS: LANGUAGES

CARTOGRAPHY: Alicia Milligan, David Banis, and Corinna Kimball-Brown

TEXT: Hunter Shobe

RESEARCH: Corinna Kimball-Brown and David Banis

KEY DATA SOURCES:

US Census Bureau. "2015 American Community Survey: Five-Year Estimates." Factfinder.census.gov.

US Census Bureau. "Language Spoken at Home by Ability to Speak English for the Population 5 Years and Over: 2017 American Community Survey." Accessed August 25, 2019. Factfinder.census.gov.

CHINATOWNS AND JAPANTOWNS

CARTOGRAPHY: David Banis

TEXT: Sarah Mercurio and Hunter Shobe

RESEARCH: Sarah Mercurio

KEY DATA SOURCES:

Dowsett, Libby. "When Portland Had the Largest Japantown in Oregon." *Street Roots News*, January 11, 2019. News. StreetRoots.org.

Hou, Jeff, et al. "Community Open Space Initiative: Chinatown, Japantown, Little Saigon—International District, Seattle." Department of Landscape Architecture, University of Washington, Community Design Studio Report, 2005. Courses.Washington.edu.

Hyatt-Evanson, Tania. "Chinese Vegetable Gardens, Portland." Oregon History Project, 2002. Accessed September 1, 2019. OregonHistoryProject.org.

Japan Center Garage Corporation. "The San Francisco Japantown History Walk." The Japantown Merchant Association, 2014. SFJapantown.org.

National Park Service. "Seattle Chinatown Historic District." Accessed September 1, 2019. Nps.gov.

Sakamoto, Henry. "Japantown, Portland (Nihonmachi)." The Oregon Encyclopedia. Accessed August 24, 2019.

The Japantown Task Force. *Images of America: San Francisco's Japantown*. Charleston: Arcadia Publishing, 2005.

Wong, Marie Rose. *Sweet Cakes, Long Journey: The Chinatowns of Portland, Oregon*. Seattle: University of Washington Press, 2004.

FOR WHERE THE BELL TOLLS

CARTOGRAPHY: Alicia Milligan and David Banis

TEXT: Hunter Shobe, Julie Wilcox, and David Banis

RESEARCH: Julie Wilcox, David Banis, Zuriel van Belle, Hunter Shobe, Sachi Arakawa, and Kristin Sellers

KEY DATA SOURCES:

Berner, Alan. "Seattle Cathedral Bells Strike Spiritual Tones." *Seattle Times*, October 12, 2017.

First Congregational United Church of Christ. "Our Church Building." Accessed August 31, 2019. UCCPortland.org.

Garofoli, Joe, and Gerald D. Adams. "Ringing of the Bellicose: Neighbor Taking North Beach Church to Court Over Noise." SFGate.com, November 21, 2003. Accessed September 2, 2019.

Korn, Peter. "For Whom Does This Bell Toll?" *Portland Tribune*, May 15, 2014. Accessed August 31, 2019.

Saint Mark's Episcopal Cathedral. "History." Accessed August 31, 2019. SaintMarks.org.

Saints Peter and Paul Church. "History." Accessed August 31, 2019. Parish.SSPeterPaulSF.org.

IV: COMMERCE

TRADE-OFFS

CARTOGRAPHY: David Banis

TEXT: Hunter Shobe and David Banis

RESEARCH: David Banis and Hunter Shobe

SPECIAL THANKS TO: Eric Memmot

KEY DATA SOURCES:

Northwest Seaport Alliance. Accessed August 10, 2019. NWSeaportAlliance.com.

Port of Portland. "World Port Source." Accessed August 10, 2019. WorldPortSource.com.

———. "Port of Portland Remains No. 1 for Auto Exports on US West Coast." March 12, 2018. Accessed August 10, 2019. PortofPortland.com.

Port of San Francisco. Accessed August 10, 2019. SFPort.com.

Port of Seattle. Accessed August 10, 2019. PortSeattle.org.

San Francisco International Airport. "Airport Commission." Accessed August 10, 2019. FlySFO.com.

US Trade Numbers. "Ports." Accessed August 10, 2019. USTradeNumbers.com.

IS THERE GAS IN THE CAR?

CARTOGRAPHY: Kelly Neeley-Brown, Liam Neeley-Brown, and David Banis

TEXT: Sachi Arakawa and Hunter Shobe

RESEARCH: Olivia Boucher, Sachi Arakawa, and David Banis

KEY DATA SOURCES:

Bay Area News Group. "Queen of the Road: Why Arco Gas Is So Cheap, and Why Some Highway Signs Confuse." *East Bay Times*, December 15, 2009. Accessed August 10, 2019.

Cameron, Jim. "Gasoline Zone Pricing—Explained and Examined." *New Haven Register*, December 17, 2017. Accessed August 10, 2019.

Douglass, Elizabeth, and Gary Cohn. "Zones of Contention in Gasoline Pricing." *Los Angeles Times*, June 19, 2005. Accessed August 10, 2019.

Garcia, Ahiza. "Why Gas Prices in California Are So High." CNN.com, February 9, 2016. Accessed August 10, 2019.

GasBuddy. Accessed September 1, 2019. GasBuddy.com.

NACS. "Why Gas Prices Vary around the Country or the Corner." March 11, 2019. Accessed August 10, 2019. Convenience.org.

US Energy Information Administration. "Gasoline Explained: Regional Gasoline Price Differences." Accessed August 10, 2019. Eia.gov.

Yantis, John. "Mom-and-Pop Gas Stations Feeling Squeeze from Major Oil Companies." *East Valley Tribune*, March 1, 2003. Accessed August 10, 2019.

HOW GREEN IS MY ENERGY?
CARTOGRAPHY: Randy Morris and David Banis
GRAPHICS: Chauncey Morris
TEXT: Hunter Shobe and David Banis
RESEARCH: David Banis and Hunter Shobe
KEY DATA SOURCES:

National Renewable Energy Laboratory (NREL). "The Geography of Green Power (2017 Data)." Accessed September 25, 2019. Nrel.gov.

Pacific Gas and Electric. "Annual Reports and Proxy Statement: 2018." Accessed September 25, 2019. Investor .PGECorp.com.

———. "Exploring Clean Energy Solutions." Accessed September 25, 2019. PGE.com.

———. "Annual Reports: 2018." Accessed September 25, 2019. Investor.PGECorp.com.

Portland General Electric. "How We Generate Electricity." Accessed September 25, 2019. PortlandGeneral.com.

Rogers, Paul. "Should Big Dams Count as Renewable Energy? California Democrats Divided." *Mercury News*, May 26, 2019. Accessed September 25, 2019.

Seattle City Light. "Annual Reports: 2018." Accessed September 25, 2019. Seattle.gov.

———. "Power Mix: How Our Electricity is Generated." Accessed September 25, 2019. Seattle.gov.

St. John, Jeff. "San Francisco Offers $2.5B to Take Over Its Share of PG&E's Grid." Clean Power Exchange, September 9, 2019. Accessed September 25, 2019. CleanPowerExchange.org.

LIVES AND LIVELIHOODS AT RISK: COVID-19
CARTOGRAPHY: Lauren McKinney-Wise and David Banis
GRAPHICS: Randy Morris
TEXT: Hunter Shobe and David Banis
RESEARCH: David Banis, Hunter Shobe, and Randy Morris
KEY DATA SOURCES:

Balk, Gene. "Seattle May Have Lowest Rate of COVID-19 Cases among Major US Cities." *Seattle Times*, October 5, 2020. Accessed December 3, 2020.

Johns Hopkins University Center for Systems Science and Engineering. "COVID-19 Dashboard by the Center for Systems Science and Engineering (CSSE) at Johns Hopkins University (JHU)." Accessed November 25, 2020. Systems .JHU.edu.

King County Public Health Department. "COVID-19 Data Dashboard." Accessed November 23, 2020. Kingcounty.gov.

Office of Governor Gavin Newsome. "California, Oregon & Washington Announce Western States Pact." April 13, 2020. Accessed December 3, 2020. Gov.ca.gov.

Oregon Health Authority. "COVID-19 Updates." Accessed November 25, 2020. Oregon.gov.

San Francisco Open Data. "COVID-19 Data and Reports." Accessed November 28, 2020. DataSF.org.

US Bureau of Labor Statistics. "Data Finder." Beta.bls.gov.

NONSTOP HUB HOP
CARTOGRAPHY: Geoff Gibson
GRAPHICS: Geoff Gibson
TEXT: Geoff Gibson
RESEARCH: Geoff Gibson
KEY DATA SOURCES:

FlightSphere. Accessed September 10, 2019. FlightSphere.com

ANALOG: BACKLASH TO DIGITAL
CARTOGRAPHY: Stephanie Sun and Zuriel van Belle
GRAPHICS: Stephanie Sun
TEXT: Hunter Shobe and Stephanie Sun
RESEARCH: Julie Wilcox and Hunter Shobe
KEY DATA SOURCES:

Harding, Colin. "Celluloid and Photography, Part 2: The Development Roll Film." *Science+Media Museum*, November 10, 2012. Accessed August 10, 2019. Blog .ScienceandMediaMuseum.org.uk.

Polt, Richard. *The Typewriter Revolution: A Typist's Companion for the Twenty-First Century.* Woodstock, Vermont: Countryman Press, 2015.

Yellow Pages. "Portland." 1987.

———. "San Francisco." 1987.

———. "Seattle." 1987.

ANALOG IN A DIGITAL WORLD
CARTOGRAPHY: Zuriel van Belle
TEXT: Hunter Shobe
RESEARCH: Julie Wilcox and Zuriel van Belle

GOTTA GO TO WORK
CARTOGRAPHY: Justin Sherrill
TEXT: Hunter Shobe
RESEARCH: Justin Sherrill, Randy Morris, and David Banis
KEY DATA SOURCES:

Fortune 500. 2019. Accessed September 25, 2019. Fortune.com.

Kish, Matthew. "Nike Joins Fortune 100, 2 Other Oregon Companies Make Fortune 500." *Portland Business Journal*, June 6, 2016. Accessed September 25, 2019.

McKenzie, Brian. "Modes Less Traveled: Commuting by Bicycle and Walking in the United States, 2008–2012." American Community Survey Reports, ACS-25. Washington, DC: US Census Bureau, May 8, 2014. Census.gov.

US Census Bureau. Longitudinal Employer-Household Dynamics: Origin-Destination Employment Statistics, 2017. Lehd.ces.census.gov.

IN THE WEEDS
CARTOGRAPHY: Kelly Neeley-Brown and Zuriel van Belle
GRAPHICS: Stephanie Sun
TEXT: Hunter Shobe
RESEARCH: Zuriel van Belle and Hunter Shobe
KEY DATA SOURCES:
Anderson, Ted. "Cannabis Price Collapse Putting Billions in Consumers' Pockets." Leafly, October 18, 2018. Accessed August 10, 2019.
McCarthy, Niall. "Which States Made the Most Tax Revenue from Marijuana in 2018?" *Forbes*, March 26, 2019. Accessed August 10, 2019.
McGreevy, Patrick. "California Might Triple the Number of Marijuana Shops Across the State." *Los Angeles Times*, May 13, 2019. Accessed August 10, 2019.
Oregon Liquor Control Commission. "Recreational Marijuana." Oregon.gov/olcc.
San Francisco Office of Cannabis. "Permitted Cannabis Retail Locations." Officeofcannabis.sfgov.org.
Washington State Liquor and Cannabis Board. "Licensed Businesses Dataset." Data.lcb.wa.gov.

V: POPULAR CULTURE

THE THRILL IS ALMOST GONE
CARTOGRAPHY: Christina Friedle
GRAPHICS: Maggie Burant and Christina Friedle
TEXT: Christina Friedle and Hunter Shobe
RESEARCH: Christina Friedle and Hunter Shobe
PHOTOGRAPHY:
Shobe, Hunter. *Oaks Amusement Park*. Portland, 2019.
KEY DATA SOURCES:
Chris6D. "The Wonders of Coney Island." Medium, June 3, 2018. Accessed August 31, 2019.
Hartlaub, Peter. "SF's Amusement Parks: Let's Take a Ride through History." *San Francisco Chronicle*, July 18, 2015.
PDXHistory. "Amusement Parks in America." Accessed August 12, 2019.
——. "Whitneys Playland at-the-Beach." Accessed August 12, 2019.
Portland Parks and Recreation. "Lotus Isle Park." Accessed August 12, 2019. Portlandoregon.gov.
Roller Coasters of the Pacific Northwest. "Defunct Coasters." Accessed August 12, 2019. RollerCoastersofthePacificNW.com.
Samuelson, Dale, with Wendy Yegoiants. *The American Amusement Park*. St. Paul, MN: MBI Publishing Company, 2001.
Seling, Megan. "Dis-Enchanted Forest: The Amusement Park at Seattle Center Is the Emptiest Place on Earth." *Stranger*, August 25, 2005. Accessed August 12, 2019.
Wyrsch, Tom, dir. *Remembering Playland at the Beach*. Garfield Lane Productions, 2010.

JAPAN + FOOD
CARTOGRAPHY: Zuriel van Belle
GRAPHICS: Zuriel van Belle
TEXT: Hunter Shobe
RESEARCH: Zuriel van Belle and Hunter Shobe
PHOTOGRAPHY:
Shobe, Hunter. *Toshihiro Kasahara, Innovator of Seattle-Style Teriyaki*. Mill Creek, 2018.
——. *Yataimura Maru*. Portland, 2019.
KEY DATA SOURCES:
Issenberg, Sasha. *The Sushi Economy: Globalization and the Making of a Modern Delicacy*. New York: Gotham Books, 2007.
Kauffman, Jonathan. "How Teriyaki Became Seattle's Own Fast-Food Phenomenon: And What the Immigrant-Fueled Dish Tells Us about Our Culture." *Seattle Weekly*, August 14, 2007.
Tomky, Naomi. "The Slow and Sad Death of Seattle's Iconic Teriyaki Scene." *Thrillist*, August 23, 2016.

THE ART OF THE POUR: BEER, WINE, AND SPIRITS
CARTOGRAPHY: Alicia Milligan
TEXT: Hunter Shobe and David Banis
RESEARCH: Julie Wilcox and David Banis
KEY DATA SOURCES:
BeerAdvocate. Accessed September 24, 2019. BeerAdvocate.com.
Booth, Stacy. "Seattle's Devoted Distillers." WhereTraveler, no date. Accessed September 24, 2019.
Drunken Diplomacy. "Northern California and Bay Area Distilleries." March 9, 2019. Accessed September 25, 2019.
Lander, Jess. "8 Urban Wineries Along San Francisco's Wine Trail." 7x7, April 11, 2017. Accessed September 24, 2019. 7x7.com.
McLean, Tessa. "The Ultimate Guide to the Gay Area's Urban Wineries." The Bold Italic, May 22, 2019. Accessed September 25, 2019.
PDX Urban Wineries. Accessed September 24, 2019. PDXUrbanWine.com.
Portland Food and Drink. "Your Guide to Portland Oregon Distilleries." Last updated July 10, 2019. Accessed September 24, 2019. PortlandFoodandDrink.com.
Seattle Urban Wineries. "Wine Country Just Got Urban." 2018. Accessed September 24, 2019. SeattleUrbanWineries.com.
West, Eric. "California Cider Map and Directory." Cider Guide, no date. Accessed September 24, 2019.
——. "Oregon Cider Map and Directory." Cider Guide, no date. Accessed September 24, 2019.
——. "Washington Cider Map and Directory." Cider Guide, no date. Accessed September 24, 2019.

8-BIT CITY
CARTOGRAPHY: Geoff Gibson
TEXT: Hunter Shobe
RESEARCH: Zuriel van Belle, Hunter Shobe, and David Banis

KEY DATA SOURCES:

BMI Gaming. "The History of Pinball Machines and Pintables." Accessed September 1, 2019.

———. "The History of Video Arcade Games." Accessed September 1, 2019.

City and County of San Francisco, Entertainment Commission. Sfgov.org.

City of Portland, Office of Management and Finance. Portlandoregon.gov.

City of Seattle, Department of Finance and Administrative Services. Seattle.gov.

History. "History of Snooker and Pool." Accessed September 1, 2019. History.co.uk.

Pinball Map. Accessed September 1, 2019. PinballMap.com.

STREET APPEAL: GRAFFITI

TEXT: Tiffany Conklin and Hunter Shobe

RESEARCH: Hunter Shobe, Tiffany Conklin, and Lourdes Jimenez

PHOTOGRAPHY:

Conklin, Tiffany, *Photos of TUBS*. Seattle, 2013.

Shobe, Hunter. *Photos of Wheatpaste*. Seattle, 2018.

———. *Photos of Stickers*. Portland, 2018.

———. *Photos of Stencils*. San Francisco, 2006–2008 and 2017–2018.

KEY DATA SOURCES:

Shobe, Hunter. "Graffiti as Communication and Language." *Handbook of the Changing World Language Map*, edited by S. Brunn and R. Kehrein. Abstract, 2018. Springer. Accessed August 10, 2019.

THE BUFF: GRAFFITI ABATEMENT

CARTOGRAPHY: Kelly Neely-Brown, Liam Neeley-Brown, and Zuriel van Belle

GRAPHICS: Galen Malcolm

TEXT: Hunter Shobe and Tiffany Conklin

RESEARCH: Hunter Shobe, Tiffany Conklin, Lourdes Jimenez, Liam Neeley-Brown, and Kelly Neely-Brown

PHOTOGRAPHY: Shobe, Hunter. *Buff in San Francisco*. San Francisco, 2008.

———. *Buff in Portland*. Portland, 2019.

KEY DATA SOURCES:

Shobe, Hunter, and David Banis. "Zero Graffiti for a Beautiful City: The Cultural Politics of Urban Space in San Francisco." *Urban Geography*, 34, no. 4 (2014), 586–607.

Shobe, Hunter, and Tiffany Conklin. "Geographies of Zero Tolerance: Graffiti Abatement in Portland, San Francisco and Seattle." *Professional Geographer*, 70, no. 4 (2018), 624–632.

GOING THE DISTANCE: MARATHONS

GRAPHICS: Maggie Burant

TEXT: Maggie Burant and Hunter Shobe

RESEARCH: Maggie Burant and Hunter Shobe

KEY DATA SOURCES:

Butler, Sarah Lorge. "After Turmoil, the Portland Marathon Will Go On." *Runner's World*, September 15, 2017. Accessed December 30, 2018.

Jones, Hugh. "History of the Marathon." Association of International Marathons and Distance Races. Reproduced from *The Expert's Guide to Marathon Training* (2003). Accessed December 30, 2018.

Longman, Jeré. "The Marathon's Random Route to Its Length." *New York Times*, April 20, 2012. Accessed December 30, 2018.

Lovett, Charlie. "Olympic Marathon Prologue: The Legend." MarathonGuide.com. Reproduced from *Olympic Marathon* (1997). Accessed December 30, 2018.

Mara, Erin. "History: Taking the Race Over the Bridge." San Francisco Marathon, no date. Accessed December 30, 2018. TheSFMarathon.com.

STADIUMS, PALACES, AND DOMES

CARTOGRAPHY: Geoff Gibson

GRAPHICS: Geoff Gibson

TEXT: Hunter Shobe

RESEARCH: Hunter Shobe, Geoff Gibson, and David Banis

PHOTOGRAPHY:

Edelstein, Ian. *Kingdome Implosion*. Color photograph, March 26, 2000. Seattle Municipal Archives Photograph Collection, Fleets and Facilities Department, record series 0207-01, item number 100538.

———. *Kingdome Implosion*. Color photograph, March 26, 2000. Seattle Municipal Archives Photograph Collection, Fleets and Facilities Department, record series 0207-01, item number 100486.

———. *Kingdome Implosion*. Color photograph, March 26, 2000. Seattle Municipal Archives Photograph Collection, Fleets and Facilities Department, record series 0207-01, item number 100494.

KEY DATA SOURCES:

Baker, Geoff. "Kingdome Debt to Be Retired 15 Years after Implosion." *Seattle Times*, March 26, 2015. Accessed August 10, 2019.

Carlson, Kip, and Paul Andresen. *The Portland Beavers*. Charleston: Arcadia Publishing, 2004.

Cornett, William. "Vaughn Street Park." The Oregon Encyclopedia. Accessed August 10, 2019.

Ezkanazi, David. "Back, Back, Back to the Ballparks." In *Rain Check: Baseball in the Pacific Northwest*, edited by Mark Armour, 63–75. Cleveland: Society for American Baseball Research, 2006.

———. "Wayback Machine: Seattle Struck Gold in Dugdale." SportsPressNW.com, July 5, 2011. Accessed August 10, 2019.

Franks, Joel S. *Whose Baseball? The National Pastime and Cultural Diversity in California, 1859–1941*. Lanham, Maryland: Scarecrow Press, 2001.

Hubbard, Anita Day. "Cities within the City: Volume II."
     Archive.org. Published in the *San Francisco Bulletin*.
     Accessed August 10, 2019.

"Ice Arena to Close." *Eugene Register-Guard*, May 27, 1953: 2A.

Mancuso, Jim, and Scott Petterson. *Hockey in Portland*.
     Charleston: Arcadia Publishing, 2007.

Nelson, Kevin. *The Golden Game: The Story of California Baseball*.
     San Francisco: California Historical Society Press, 2004.

Obermeyer, Jeff. "The Arenas." SeattleHockey.net, 2009.
     Accessed August 10, 2019.

Orr, Michael. "Multnomah Stadium." The Oregon
     Encyclopedia. Accessed August 10, 2019.

Raley, Dan. "A Man Named Sick Made Seattle Well Again." In
     Rain Check: Baseball in the Pacific Northwest, edited by Mark
     Armour, 56–62. Cleveland: Society for American Baseball
     Research.

Raley, Dan. *Pitchers of Beer: The Story of the Seattle Rainiers*.
     Lincoln: University of Nebraska Press, 2011.

Riess, Steven A. *City Games: The Evolution of American Urban
     Society and the Rise of Sports*. Urbana: University of Illinois
     Press, 1989.

Rose Quarter. "Moda Center." Accessed September 1, 2019.

———. "Veterans Memorial Coliseum." Accessed August 10,
     2019.

Shobe, Hunter, and Geoff Gibson. "Cascadia Rising: Soccer,
     Region and Identity." *Soccer and Society*, 18, no. 7, 953–971.

Siegelbaum, Lewis H., and Sasu Siegelbaum. "Class and
     Sport." In *The Oxford Handbook of Sports History*, edited by
     R. Edelman and W. Wilson, 429–443. New York: Oxford
     University Press, 2017.

Simons, Terry. *The Children of Vaughn: The Story of Professional
     Baseball in Portland, Oregon (1901–2010)*. Portland: Round
     Bend Press Books, 2014.

Stein, Alan J. "Sicks' Stadium." HistoryLink.org, July 15, 1999.
     Accessed August 10, 2019.

REVERT TO TYPE

CARTOGRAPHY: Hunter Shobe and Zuriel van Belle
TEXT: Hunter Shobe
SPECIAL THANKS TO: Barbara Brower for use of her Hermes
     Rocket 3000 and Matt McCormack at Ace Typewriter in
     Portland for the new ribbon and a wealth of information
     about typewriters.

NOTED PLACES

CARTOGRAPHY: Hunter Shobe, Zuriel van Belle, Jonathan
     van Belle, and David Banis
TEXT: Hunter Shobe and David Banis
RESEARCH: Hunter Shobe and David Banis
SPECIAL THANKS TO: Kezia Rasmussen

CREDITS

Page 52: Reproduced by permission from the San Francisco History Center, San Francisco Public Library. *Construction of the Embarcadero Freeway* by photographer Eddie Murphy. Black and white photograph, October 24, 1955. San Francisco Historical Photograph Collection, photo identification number AAF-0882.

Page 56: Reproduced by permission from the San Francisco History Center, San Francisco Public Library. *Workmen Digging up Graves in Odd Fellows Cemetery*. Black and white photograph, September 28, 1931, San Francisco Historical Photograph Collection, photo identification number AAD-6218.

Pages 60 and 61: Sailboat icons by Smalllike from the Noun Project, TheNounProject.com.

Pages 60 and 61: Houseboat icons by Peter van Driel from the Noun Project, TheNounProject.com.

Page 71: Reproduced by permission from the San Francisco History Center, San Francisco Public Library. *Elephant, SF Fleishhacker Zoo*. Black and white photograph, January 1, 1949. San Francisco Fleishhacker Zoo Collection, negative number 9944.

Page 86: Reproduced by permission from the San Francisco History Center, San Francisco Public Library. *Panoramic View of Earthquake and Fire Damage around Union Square, Taken from Intersection of Stockton and Post Streets, Looking Southwest toward the St. Francis Hotel*. Black and white photograph, ca. 1906. San Francisco Historical Photograph Collection, photo identification number AAC-4079.

Page 106: Reproduced by permission from Museum of History and Industry. *Group around Table at the Black and Tan, Seattle, Circa 1947* by photographer Albert J. Smith Sr. Black and white photograph, ca. 1947. Museum of History and Industry, Seattle, Al Smith Collection, image number 2014.49.010-051-0102.

Page 107: Reproduced by permission from the San Francisco History Center, San Francisco Public Library. Canterbury, Alan J. *View of Fillmore Street from Bush, Looking South*. Black and white photograph, June 19, 1964. San Francisco Historical Photograph Collection, photo identification number AAB-3649.

Page 108: Reproduced by permission from Michael Henniger. *A Photograph of Duke Ellington Performing at McElroy's Spanish Ballroom (1953)* by photographer Carl Henniger. Black and white photograph, ca. 1953. Portland, Oregon.

Page 109: Reproduced by permission from Tim Berkley. *San Francisco Jazz Artists Kim Nalley, Standing, and Tammy Hall, at the Piano, Fillmore Jazz Festival* by photographer Tim Berkley. Color photograph, ca. 2017. San Francisco, California.

Page 189: Reproduced by permission from C. Bruce Forster. *Graffiti Photos from Pirate Town* by photographer C. Bruce Forster. Color photograph, date unknown. Portland, Oregon.

Page 189: Reproduced by permission from Michael Endicott. *Temple of the Urban Cave Painters 3* by photographer Michael Endicott. Color photograph, July 21, 2005. Portland, Oregon.

Page 189: Reproduced by permission from Rachel Escoto. *Treasure Island Photographs* by photographer Rachel Escoto. Color photographs, ca. 2014. San Francisco, California.

Page 196: Reproduced by permission from Donald Nelson and the Vince Pesky Collection/Nelson Archives. *A Photograph of Vaughn Street Stadium circa 1931*. Black and white photograph, ca. 1931. Portland, Oregon.

Page 199: Reproduced by permission from the San Francisco History Center, San Francisco Public Library. *View of Kezar Stadium and Golden Gate Park* by photographer Hosea Blair. Black and white photograph, ca. 1925–1958. San Francisco Historical Photograph Collection, photo identification number AAC-5213.

Printed in Korea

SASQUATCH BOOKS with colophon is a registered trademark
of Penguin Random House LLC

25 24 23 22 21   9 8 7 6 5 4 3 2 1

Editors: Gary Luke and Jen Worick
Production editor: Jill Saginario
Designer: Anna Goldstein
Production designer: Alison Keefe

Library of Congress Cataloging-in-Publication Data:
Names: Banis, David, cartographer, author. | Shobe, Hunter,
   cartographer, author. | Belle, Zuriel van, cartographer, author. |
   Gibson, Geoff (Transportation planner), cartographer, contributor. |
   Arakawa, Sachi, cartographer, contributor.
Title: Upper left cities : a cultural atlas of San Francisco, Portland, and
   Seattle / David Banis and Hunter Shobe with Zuriel van Belle ; lead
   contributors, Geoff Gibson, Sachi Arakawa.
Description: Seattle : Sasquatch Books, [2021] | "Copyright © 2021 by
   Hunter Shobe and David Banis." | Includes bibliographical references.
Identifiers: LCCN 2019047467 | ISBN 9781632171825 (hardcover)
Classification: LCC G1489.S4E6 B3 2020 | DDC 912.741/91--dc23
LC record available at https://lccn.loc.gov/2019047467

ISBN: 978-1-63217-182-5

Sasquatch Books
1904 Third Avenue, Suite 710
Seattle, WA 98101

SasquatchBooks.com